Norbert Gust

Verbesserung der Signalauswertung für die Ultraschallmikroskopie

Improvement of Signal Analysis for the Ultrasonic Microscopy

TUDpress

DRESDNER BEITRÄGE ZUR ZERSTÖRUNGSFREIEN PRÜFTECHNIK
Herausgegeben von K.-J. Wolter, N. Meyendorf, H. Heuer
Band 2

Norbert Gust

Verbesserung der Signalauswertung für die Ultraschallmikroskopie

Improvement of Signal Analysis for the Ultrasonic Microscopy

TUDpress
2011

Die vorliegende Arbeit wurde am 23. Februar 2010 als Dissertation an der Fakultät Elektrotechnik und Informationstechnik der Technischen Universität Dresden eingereicht und am 21. September 2010 erfolgreich verteidigt.

Promotionskommission:

Vorsitzender:	Prof. Dr.-Ing. habil. Czarske
Gutachter:	Prof. Dr. rer. nat. et Ing. habil. Elfgard Kühnicke
	Prof. Dr. habil. Norbert Meyendorf

Bibliografische Information der Deutschen Nationalbibliothek

Die Deutsche Nationalbibliothek verzeichnet diese Publikation in der Deutschen Nationalbibliografie; detaillierte bibliografische Daten sind im Internet über http://dnb.d-nb.de abrufbar.

Bibliographic information published by the Deutsche Nationalbibliothek

The Deutsche Nationalbibliothek lists this publication in the Deutsche Nationalbibliografie; detailed bibliographic data are available in the Internet at http://dnb.d-nb.de.

ISBN 978-3-942710-22-0

Verlag der Wissenschaften GmbH
Bergstr. 70 | D-01069 Dresden
Tel.: 0351/47 96 97 20 | Fax: 0351/47 96 08 19
http://www.tudpress.de

Kurzfassung

Diese Dissertation befasst sich mit der Verbesserung der Signalauswertung für die Ultraschallmikroskopie. Scannende Ultraschallmikroskope nutzen Schallköpfe mit einer Mittenfrequenz von 10 - 300 MHz und stellen die mit dem Impuls-Echo-Verfahren gewonnenen Reflexionssignale in A-, B- oder C-Scans dar. Ausgewertet wird in der Regel jeweils die erste Reflexion von einer Grenzschicht, deren Amplitude oder Laufzeit mit Falschfarben abgebildet wird. Die Amplitude enthält hauptsächlich Informationen über die Materialübergänge, wodurch z.B. Delaminationen oder Einschlüsse erkannt werden können. Die Laufzeit der Reflexionen wird von dem Weg und der Schallgeschwindigkeit im Material bestimmt. Mit der Laufzeit in einzelnen Schichten können z.B. Dickenschwankungen oder Materialinhomogenitäten untersucht werden. Sind Weg und Schallgeschwindigkeit unbekannt, kann allein aus der Laufzeit keine der Größen bestimmt werden. Häufig werden in diesem Fall mittlere Annahmen für die Schallgeschwindigkeit genutzt, um die Tiefenlage von Grenzschichten fehlerbehaftet darzustellen.

Eine große Herausforderung stellen insbesondere Proben mit vielen und/oder dünnen Schichten dar. Hier ist aufgrund der vielen Materialübergänge die reflektierte Schallenergie von tieferen Schichten sehr gering, so dass deren Echos eine Amplitude deutlich unterhalb der des Signalrauschens besitzen. Schichtdicken unterhalb der Impulslänge bewirken außerdem eine Überlagerung der empfangenen Echos, die dann nicht mehr getrennt ausgewertet oder dargestellt werden können. Viele Schichten führen des Weiteren zu einer exponentiell wachsenden Anzahl von Mehrfachreflexionen, die die ersten Reflexionen tieferer Grenzschichten überlagern und eine eindeutige Zuordnung erheblich erschweren. Andererseits ist es bei einer großen Anzahl überlagerter Mehrfachreflexionen durch die Aufsummierung deren Amplituden möglich, auch Informationen von tiefen Grenzschichten mit ausreichendem Signal-Rausch-Abstand zu erhalten. Eine Auswertung der Mehrfachreflexionen mit klassischen Ansätzen wie Laufzeit- und Spitzenwertdarstellung führt zu falschen Aussagen, da Mehrfachreflexionen unterschiedliche Laufwege haben und jeder Laufweg länger ist als der des primären Echos.

Für die Verbesserung der Signalauswertung wurde ein iteratives Rekonstruktionssystem entworfen, das durch vergleichende Simulation erheblich mehr und genauere Informationen aus den gemessenen Ultraschallsignalen gewinnt. Dabei werden die Ergebnisse der Ultraschallsimulation der zu untersuchenden Probe mit den Ultraschallmessungen verglichen, und aus den Abweichungen nach jeder Iteration genauere Simulationsparameter bestimmt. Nach mehreren Iterationen können auf diese Weise die Materialparameter und der Aufbau der Probe bestimmt werden.

Eine grundlegende Voraussetzung für den iterativen Rekonstruktionsansatz ist eine Ultraschallsimulation, die die gemessenen Ultraschallsignale mit allen wesentlichen Effekten nachbildet. Die Simulation sollte für den praktischen Nutzen sehr schnell sein. Dazu wurde ein Algorithmus entwickelt, der das Schallfeld in ebene Wellen zerlegt und deren Ausbreitung mit Strahlverfolgung simuliert. Die daraus gewonnene Impulsantwort wird anschließend mit Referenzsignalen gefaltet und ergibt das simulierte Ultraschallsignal. Um auch komplexere Proben mit Strahlverfolgung simulieren zu können, wurde ein optimierter Algorithmus entwickelt, der den mit der Anzahl der Schichten exponentiell wachsenden Rechenaufwand zu einer linearen Abhängigkeit überführt. Die damit erzielten Ergebnisse zeigen eine hohe Wirklichkeitstreue bei einer sehr geringen Rechenzeit von weniger als eine Sekunde pro Reflexionssignal bei einer Probe mit 19 Schichten.

Für eine erfolgreiche Rekonstruktion ist es notwendig, die grundlegende Struktur einer Probe vorher zu bestimmen. Dazu gehören die Dicken und die groben Materialparameter aller zu rekonstruierenden Schichten der Probe. In der Arbeit werden mehrere Ansätze vorgestellt, mit denen je nach Anforderungen und Vorwissen Dichte, elastische Module, Schallgeschwindigkeit und Dämpfung bestimmt werden können. Da es jedoch nicht immer möglich ist, die einzelnen Schichtdicken direkt zu bestimmen, wurde ein neues Verfahren entwickelt, das die gleichzeitige Bestimmung von Schallgeschwindigkeiten und Schichtdicken allein aus Laufzeitmessungen ermöglicht. Dazu muss auf jede Grenzschicht der Probe einzeln fokussiert werden. Aus den Fokuspositionen und den Laufzeitdifferenzen der jeweiligen Schicht kann nun die Dicke bestimmt werden. Mit einer Kalibrierung auf eine Referenzmessung lässt sich mit dem neuen Verfahren eine Genauigkeit von etwa 99% erzielen.

Ist die grundlegende Struktur einer Probe bekannt, können anhand der Reflexionsparameter von Simulation und Messung genauere Materialparameter bestimmt werden. Um die Reflexionsparameter aus den Ultraschallsignalen möglichst genau und auch bei Überlagerung der Echos zu bestimmen, wurde ein Entfaltungsalgorithmus entwickelt, der wiederum die bisher bekannte Struktur der Probe nutzt, um bessere Impulsformvorhersagen zu treffen. Die Simulation, Entfaltung und Parameteranpassung wird so oft wiederholt, bis eine ausreichende Genauigkeit erreicht ist. Bei gängigen Komponenten der Halbleiterindustrie lassen sich die Reflexionsparameter für bis zu drei Grenzschichten durch Entfaltung gut bestimmen. Bei tieferen Grenzschichten werden die primären Reflexionen von Mehrfachreflexionen zu stark überlagert, wodurch eine sichere Bestimmung der Reflexionsparameter nicht mehr möglich ist. Es ist jedoch möglich alle Parameter der Simulation zu optimieren, indem die normierte Korrelation zwischen Simulation und Messung ausgewertet wird. Auf diese Weise lassen sich auch die genauen Parameter tiefer liegender Schichten bestimmen, da mit dieser Methode auch Informationen aus den überlagerten Mehrfachreflexionen gewonnen wird.

Mit der genauen Kenntnis der idealen Struktur der Probe ist es jetzt möglich, unterschiedliche Defekte zu simulieren. Durch Vergleich aller gemessenen Ultraschallsignale mit den simulierten möglichen Probendefekten kann jedem Punkt der realen Probe eine Wahrscheinlichkeit für das Vorliegen eines bestimmten Defektes zugeordnet werden. Das so erhaltene Klassifikationsergebnis lässt sich dreidimensional als Volumen darstellen, wobei für jeden Volumenpunkt der Probe das Material mit den entsprechenden Parametern nun bekannt ist.

Die in dieser Arbeit entwickelten Algorithmen lassen sich auch für bessere Ergebnisse bei der klassischen Ultraschallanalyse nutzen. Amplituden und Laufzeiten von Echos können durch die Entfaltung um mehr als eine Größenordnung genauer bestimmt und bei Überlagerung getrennt werden. Dadurch verringert sich unter anderem das Rauschen, was zu besserer Bildqualität und Interpretierbarkeit führt. Mit Hilfe der realitätsnahen Simulation ist es jetzt außerdem möglich, Schallausbreitungseffekte besser zu verstehen, die optimale Messanordnung zu finden und die erhaltenen Ergebnisse sicherer und besser zu interpretieren.

Aus den Erkenntnissen dieser Dissertation wurde ein Ultraschall-Analyseprogramm entwickelt, das diese komplexen Funktionen auf einer gut bedienbaren Oberfläche bereitstellt und bereits praktisch genutzt wird. Im Ausblick werden zahlreiche weitere Ideen vorgestellt, mit denen die Ultraschallanalyse weiter verbessert werden kann, um noch mehr und genauere Informationen zu erhalten.

Abstract

This thesis develops methods for improving signal analysis in ultrasonic microscopy. Scanning ultrasonic microscopes are equipped with transducers with a center frequency ranging from 10 MHz up to 300 MHz. The ultrasonic reflection signals are obtained with the impulse-echo method, where the data is displayed in A-, B- or C-Scans. Usually the time or the amplitude of the first reflection from an interface is evaluated and displayed. The amplitude mainly contains information about the material interfaces that can be used to detect delaminations or inclusions. The time-of-flight is determined by the sound path and the sound velocities, which makes it very sensitive to thickness or material variations. If sound velocity and sound path are unknown, it is not possible to determine one of the values only from the time-of-flight. Usually, average sound velocity assumptions are used in this case to display the depth of interfaces with some error.

One major challenge are specimens with many and/or thin layers. Numerous material interfaces reduce the reflection energy of deeper layers, and their reflection amplitudes become lower than the signal noise. Moreover, layers thinner than pulse length lead to pulse superimposition, so they cannot be analyzed or displayed independently. Additionally, numerous layers lead to an exponential growth of the number of multiple reflections. They superimpose the reflections of deeper layers, and make it almost impossible to identify single echoes. On the other hand, the great number of multiple reflections can provide enough energy for a measurement with sufficient signal-to-noise ratio, so information from deeper layers can be obtained. The evaluation of multiple reflections with classical approaches like time-of-flight and peak value images leads to erroneous conclusions, since multiple reflections have different sound paths, where each is longer than the path of the primary echo.

For the improvement of the signal analysis for ultrasonic microscopy, an iterative reconstruction system has been developed. It gains increasingly precise information from ultrasonic measurements by comparing simulations with the measurement. From the differences between simulation and measurement more precise simulation parameters are determined for the next iteration. After several iterations, this process delivers the material characteristics and the structure of the specimen.

One basic requirement for the iterative reconstruction approach is an ultrasonic simulation which models the measured ultrasonic signals with all relevant effects. For practical purpose, this simulation should be very fast. To that end, an algorithm has been developed which decomposes the sound field into plane waves, and simulates their propagation with ray tracing. The resulting impulse response is convoluted with reference signals, and the simulated ultrasonic signal is obtained. For the simulation of complex specimens an optimized algorithm has been developed, which reduces the exponential effort of ray-tracing to a linear dependency of the number of layers. The results obtained with this algorithm are very close to reality, with a computation time under one second for one reflection signal of a 19-layer specimen.

A successful reconstruction requires prior knowledge of the specimen's basic structure. This includes the layer thicknesses, and the basic material characteristics. Depending on previous knowledge and restrictions, several approaches have been developed to estimate density, elastic modules, sound velocity, and attenuation. Since it is not always possible to measure the thickness of the single layers directly, a new method has been developed to allow the simultaneous estimation of the sound velocities and the thicknesses from time-of-flight measurements. For that, each interface of the specimen is focussed separately. From the focus positions and the time-of-flight of each layer, the thicknesses and sound velocities are determined. The new method can achieve a precision of 99% with a calibration on a reference measurement.

Once the basic structure of a specimen is known, the reflection parameters from simulation and measurement can be used to determine more accurate material parameters. For a precise determination of the reflection parameters from the ultrasonic signals even under pulse superimposition, a deconvolution algorithm has been developed. It uses the known structure for a better prediction of the impulse shapes. All steps - the simulation, deconvolution and parameter adaptation - are repeated until sufficient precision if achieved. For common semi-conductor components, the reflection parameters of the first three interfaces can be easily determined. However, a proper separation is no longer possible for highly superimposed reflections of deeper interfaces. Here, the normalized correlation between simulation and measurement can be used to optimize all simulation parameters. This method enables the precise determination of parameters from deeper layers, since information is also obtained from superimposed multiple reflections.

With the precise knowledge of the ideal structure of the specimen it is now possible to simulate different defects of the specimen. By comparing all measured ultrasonic signals with the simulated possible defects, a probability for a certain defect can be determined for each point in the specimen. With a classification according to the probabilities, the specimen can be displayed as a three-dimensional volume, where for each volume point the material with its characteristics is known.

The algorithms developed in this thesis can also be used to improve the results of the classical ultrasonic analysis. With the deconvolution, amplitudes and times-of-flight can be determined more than 10 times more precisely than with classical peak detection. This reduces noise and provides better quality and ease of interpretation. The deconvolution also helps to separate superimposed reflections from thin layers. With a simulation this close to reality, it is possible to perform feasibility investigations, to find the optimal measurement setup, and to understand and interpret results with more accuracy and confidence.

The gained knowledge of this thesis has been used for the development of an ultrasonic analysis software, which provides the complex analysis functions under an easy to handle user interface. This program is already used practically. Numerous ideas for future developments are presented in the perspectives. They can help to improve the ultrasonic signal analysis even more to obtain more and better information.

Contents

Kurzfassung ..... **V**

Abstract ..... **VIII**

List ob abbreviations ..... **XIII**

1 Introduction **1**
- 1.1 Motivation 2
- 1.2 System theoretical description 3
- 1.3 Structure of the thesis 6

2 Sound field **8**
- 2.1 Sound field measurement 8
- 2.2 Sound field modeling 11
  - 2.2.1 Reflection and transmission coefficients 11
  - 2.2.2 Sound field modeling with plane waves 13
  - 2.2.3 Generalized sound field position 19
- 2.3 Receiving transducer signal 20
  - 2.3.1 Calculation of the transducer signal from the sound field 20
  - 2.3.2 Received signal amplitude 21
  - 2.3.3 Measurement of reference signals 24

3 Ultrasonic Simulation **27**
- 3.1 State of the art 27
- 3.2 Simulation approach 28
  - 3.2.1 Sound field measurement based simulation 28
  - 3.2.2 Reference signal based simulation 30
- 3.3 Determination of the impulse response 31
  - 3.3.1 1D ray-trace algorithm 31
  - 3.3.2 2D ray-trace algorithm 33
  - 3.3.3 Complexity reduction – optimizations 35

4 Deconvolution – Determination of reflection parameters **38**
- 4.1 State of the art 39
  - 4.1.1 Decomposition techniques 39
  - 4.1.2 Deconvolution 41
- 4.2 Analytic signal investigations for deconvolution 42
- 4.3 Single reference pulse deconvolution 44
- 4.4 Multi-pulse deconvolution 47
  - 4.4.1 Homogeneous multi-pulse deconvolution 48
  - 4.4.2 Multi-pulse deconvolution with simulated GSP profile 49

5 Reconstruction **50**
- 5.1 State of the art 50
- 5.2 Reconstruction approach 51
- 5.3 Direct material parameter estimation 52
  - 5.3.1 Sound velocities and layer thickness 52
  - 5.3.2 Density, elastic modules and acoustic attenuation 54
- 5.4 Iterative material parameter determination of a single layer 56
- 5.5 Reconstruction of complex specimens 60
  - 5.5.1 Material characterization of multiple layers 60

5.5.2 Iterative simulation parameter optimization with correlation ... 62
5.5.3 Pattern recognition reconstruction of specimens with known base structure. 66

6 Applications and results ... 71

6.1 Analysis of stacked components ... 71
6.2 Time-of-flight and material analysis ... 74

7 Conclusions and perspectives ... 78

References ... 82

Figures ... 86

Tables ... 88

Appendix ... 89

Acknowledgments ... 100

Danksagung ... 101

Curriculum vitae ... 102

List ob abbreviations

Abbreviation	Unit	Description
1D, 2D, 3D		One-, two-, three-dimensional
A-Scan		Amplitude Scan: Reflection or transmission signal at a fixed transducer location (x, y, z); sound propagation in z direction; all A-Scans in this thesis have a normalized amplitude
B-Scan		Brightness Scan: The A-Scan data along a line in the x-y-plane is displayed with color coded amplitudes (historically: bright = high amplitude, dark = zero; in this thesis: normalized amplitudes with dark = positive amplitude, white = zero, red = negative amplitude); the abscissa of the image represents the location on the line and the ordinate represents the time-of-flight (z-axis sound velocity proportional)
C-Scan		2D x-y-scan with fixed z-position: It usually displays the color coded (see B-Scan) maximum amplitude within a fixed time gate; there can also be other evaluation methods like averaging, energy determination, correlation or time-of-flight analysis
Z-Scan		1D z-scan or depth-scan: The A-Scan data is recorded at a fixed x-y-location for several z-positions. The displayed image data in this thesis is triggered on the surface echo of the specimen, where the abscissa of the image represents the time-of-flight since the surface trigger (z-axis sound velocity proportional) and the ordinate represents the surface echo time-of-flight since pulse excitation (z-position of the transducer sound velocity proportional).
Echo, Reflection		The received reflection from a single material interface. Different reflection modes are received as separate reflections. All reflections from a specimen are contained in the reflection signal.
erf(x)		Error function: $\operatorname{erf}(x) = \frac{2}{\sqrt{\pi}} \int_0^x e^{-y^2} dy$
exp(x)		Exponential function with basis e: $\exp(x) = e^x$
FFT		Fast Fourier Transformation (inverse: FFT^{-1})
sign(x)		Sign function: $\operatorname{sign}(x) = x / \lvert x \rvert$
SNR	dB	Signal-to-noise ratio
TOF	s	Time-of-flight since pulse excitation
GSP	m^2 / s	Generalized Sound field Position (chapter 2.2.3)
c_P, c_S	m / s	Pressure- and shear-wave sound velocity
ρ	kg / m^3	Mass density
a	dB / m	Average ultrasonic attenuation

1 Introduction

Ultrasonic microscopes are widely used for nondestructive testing and evaluation. The commonly used spherical focused transducers have a center frequency from 10 MHz up to 300 MHz and a resolution down to 7 µm depending on the materials. Especially for electronic components they are used to get information with respect to delaminations within the package due to production, operation or reliability testing. The tendency towards smaller and more complex structures of components leads to signal superposition of different interfaces, to a decreased signal to noise ratio and consequently to a difficult or even wrong interpretation of the measurement results by the user. Avoiding misinterpretation of failure localization as well as getting additional information on material properties, an advanced ultrasonic analysis and a better understanding of sound propagation with its effects are required.

The goal of this thesis is to extract the maximum possible information from ultrasonic signals without additional hardware improvements. For this, the whole ultrasonic analysis system is going to be investigated and each single part of it will be investigated in detail. From the knowledge of the ultrasonic system (signal, transducer and specimen) and the comparison to the measured data, more information can be obtained than usually is gained from ultrasonic inspections.

This system approach requires accurate data analysis, a realistic and efficient system description and intelligent reconstruction algorithms, to achieve quantitatively precise results within acceptable processing time. The ultrasonic data analysis is necessary to determine the reflection parameters from measured ultrasonic signals. Since those reflections can be very closely spaced or even superimposing, a deconvolution algorithm has to be used. This deconvolution algorithm should be capable to handle varying pulse shapes to compensate the different interference conditions of the received ultrasonic echoes.

The description of an ultrasonic system can be done with a sound field simulation. Realistic three-dimensional sound field calculations usually have a high computational effort, so efficient simplifications are necessary to obtain results with sufficient quality in short processing time. Comparable realistic ultrasonic simulations also require the knowledge of the original sound field from the measurement system. This information is obtained from sound field and reference signal measurements.

The evaluation of the differences between simulation and measurement delivers information about the variations of the real specimen compared to the simulated one. By adaptation of those changes, the actual specimen can be reconstructed with an iterative process. This approach provides more and better information than direct ultrasonic data analysis.

1.1 Motivation

Ultrasonic inspection techniques are evaluating the amplitude, the spectrum and the time-of-flight of single interface echoes. Problems arise when those echoes are superimposed with other echoes in specimens with multiple and/or thin layers. In many cases, a separation of the ultrasonic echoes can be helpful, because it allows independent inspection of interfaces between thin layers.

Information on the specimen's structure (interfaces) can not be found only in the first interface echo (primary echo). All multiple reflections and mode converted echoes also contain information, which is spread over the whole ultrasonic signal. The challenge is to understand and to evaluate these effects. Signal changes due to flaws in complex specimens are often detectable, but it is not possible to determine the location and the type of the flaw. The single echoes are superimposed and the time of the signal change can occur later than expected. The summation of multiple echoes might have a much higher amplitude than the primary echoes as in figure 1a. An ultrasonic simulation, as in reference [1], for predicting signal changes is highly desirable. It can help to understand propagation effects like influences of the materials on the signal, the type and the characteristics of ultrasonic echoes, changes of the signal due to specific flaws, and influences from focusing and transducer choice. A realistic simulation based on measured sound field parameters could also allow the direct comparison of the measured signals with the simulated ones. Common ultrasonic simulations in ultrasonic analysis systems are based on rudimentary assumptions and do not deliver sufficient quality for many applications. On the other hand there are very realistic scientific simulation algorithms available, which require a lot of computation time. The reconstructive ultrasonic analysis for practical use requires a fast ultrasonic simulation with sufficient quality.

Normally, the elastic material parameters of the specimen are not very well known or unknown. They are required for the ultrasonic simulation, and the precision and the characteristics of the results are very much affected by differences with the actual parameters. A determination of those parameters can be very difficult, and in many cases it is not possible to characterize one material in a structure of different materials independently. Since ultrasonic signals contain most of the information that is required for the material characterization, a reconstruction algorithm could deliver the elastic parameters and structure information.

This all requires an improvement of the ultrasonic analysis system with a complete system description. It is geared toward practical application, so every part of the analysis system must be optimized to obtain high quality results within short processing time.

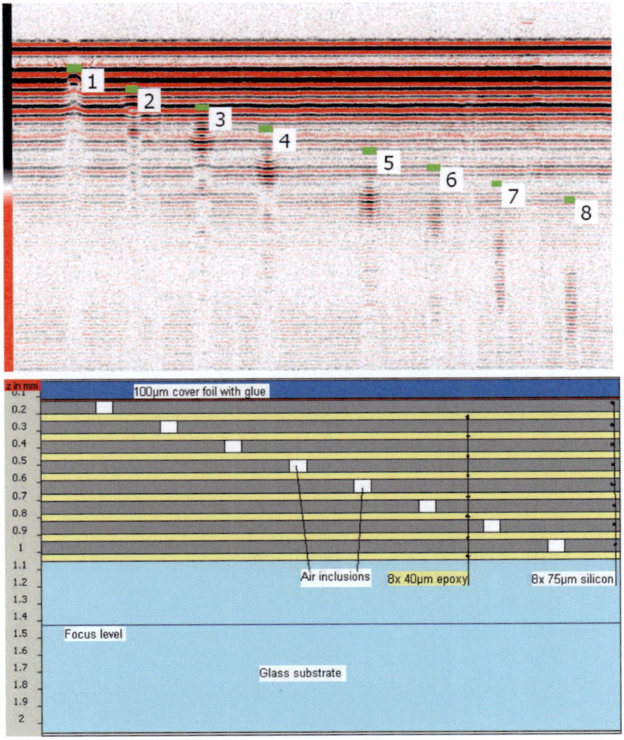

a)

b)

Figure 1: Reference specimen 3: eight-die stack with air-inclusions
a) B-Scan through the flaws of each die; green bars: position marker of primary flaw echo for each die (number); scan area: 17.5 mm x 1.1 µs; focus inside glass substrate
b) Construction scheme of reference specimen 3 (*x*-*z*-view, exact specimen description on appendix pages 97-99)

1.2 System theoretical description

A conventional ultrasonic microscope (figure 2) consists of an ultrasonic transducer which is attached to a x-y-z-scanner. This transducer is excited with an impulse generator and emits a focused wave into the coupling water. The wave is scattered, reflected and transmitted at the interfaces of the specimen and attenuated while propagating through materials. The reflected waves are received by the same transducer (and/or the transmitted part is received by a second transducer) and converted to an electrical signal (received signal). In older systems this signal is analyzed directly with analog peak detectors whereas in modern systems the signal is converted to a digital signal first and then evaluated with digital signal processing.

Under idealized conditions (no distortions, low power, etc.) linearity can be assumed and the signal path is described with a linear system. For such a system, the received signal $s_r(t)$ is determined from the time convolution ($*$ symbol) with the impulse response of the signal path and the integration over the surface of the emitting and the receiving transducer:

$$s_r(t) = I(t) * RE(t) * \int\limits_{x_1}\int\limits_{y_1}\int\limits_{x_2}\int\limits_{y_2} TD_1(t, x_1, y_1) * SP(t, x_1, y_1, x_2, y_2) * TD_2(t, x_2, y_2) dy_2 dx_2 dy_1 dx_1 \text{ ,(1)}$$

where $I(t)$ the excitation impulse signal, TD the transformation factor and the impulse response of a transducer at the coupling level z ($TD_1 = TD_2$ for an one transducer system), SP the impulse response from a point (x_1, y_1, z_1) to another point (x_2, y_2, z_2) of the specimen and RE the impulse response of the receiving unit (figure 2). For convenience, the interface level z between transducer and specimen can be in the water delay line.

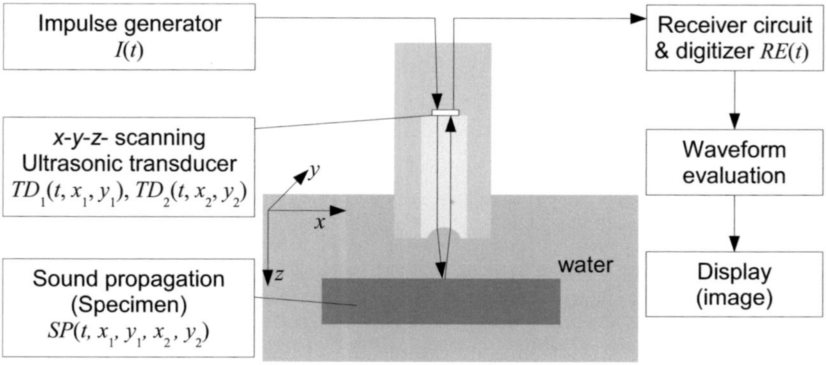

Figure 2: Conventional ultrasonic analysis scheme

The information that is used in a conventional analysis system usually is the time-of-flight and/or the maximum amplitude of primary interface echoes. It is displayed in dependence of the scanner position in an image with a gray scale or a color map. This type of analysis has no feedback loops. Errors and variations of the sub-systems are accumulated in the results. A compensation of some errors and variations is possible with reference measurements as in chapter 2.3.3. With a known reference specimen they can be used to determine reference signal amplitudes for calibration. They have to be done again if the system is changed (e.g. temperature). It requires reference specimens and additional time effort for extra measurements.

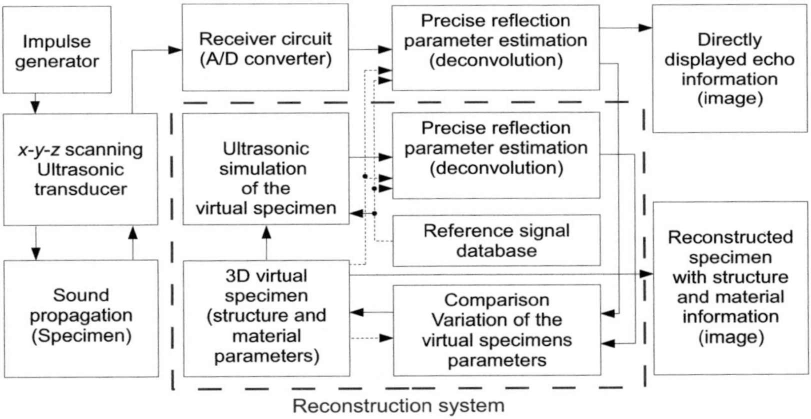

Figure 3: Ultrasonic system analysis scheme

In this thesis the conventional ultrasonic analysis system is extended with a reconstruction system (figure 3). It models the measured specimen with a simulated virtual specimen and compares the measured signals to the simulated ones. From the differences of those signals the parameters of the virtual specimen are calculated and the next iteration loop is started. With this iterative reconstruction system it is possible to extract the maximum possible information of a measured ultrasonic signal. A reference specimen or a reference measurement is not required. Changes in the system, like the temperature, can be considered in the simulation. The precision of the new analysis system only depends on the quality of the ultrasonic simulation and the accuracy of the reflection parameter estimation. Therefore, the development of a fast and realistic ultrasonic simulation and a deconvolution algorithm for the reflection parameter estimation is a major part of this thesis. This complete system approach delivers the information of the whole system and not only the reflection parameters of a single echo as in the conventional analysis.

1.3 Structure of the thesis

The general approach for the improvement of the signal analysis is described in chapter 1.2. The implementation of a reconstruction system (figure 3) into the data analysis of ultrasonic microscopes enables inverse analysis of ultrasonic data and thereby provides more and better information about the investigated specimen than data analysis without reconstruction. All relevant components of the ultrasonic signal analysis and the reconstruction system are developed for different analysis situations, to obtain optimal results with reasonable effort. The major aspects are the sound field modeling for the sound propagation in the specimen, the analysis of the ultrasonic signals and the reconstruction algorithms. This all is combined into a new ultrasonic analysis system.

The basis for the system description is the modeling of the sound field. For that, the following sound field considerations are treated in chapter 2:

- Sound field measurement as input for realistic ultrasonic simulations
- Determination of reflection and transmission coefficients at material interfaces
- Modeling of wave propagation with a plane wave decomposition model, which is well suited for efficient ultrasonic ray trace simulations
- Definition of the generalized sound field position for shape description of the sound field in layered structures
- Determination of the ultrasonic signal and it's characteristics from the sound field
- Measurement of reference signals and data processing

Based on the sound field modeling and the decomposition into plane waves, the ultrasonic simulation algorithm for the system description is developed in chapter 3. The reflection signals are determined by the convolution of the impulse response with the sound field measurement or the reference signals. The determination of the impulse response of the specimen as part of the ultrasonic simulation is done with ray trace algorithms, which are based on different assumptions, and the quality of the results is compared. Since the effort for ray trace simulations is exponential with the complexity of the specimen, simplifications and optimizations are introduced.

In chapter 4 the deconvolution of the ultrasonic signal for the precise determination of the reflection parameters is handled. The deconvolution uses a further developed iterative algorithm, which decomposes the ultrasonic signal in reference signals. Preliminarily signal investigations are done to find optimal working conditions. Additionally, different reference signal prediction models are investigated, which are suited for different analysis situations due to their stability and prediction quality.

Finally, the reconstruction approaches are treated in chapter 5. First is a description of the direct methods that determine material characteristics directly from reflection parameters. Usually, those direct methods are based on simplified assumptions. For higher precision, an iterative method for single layer reconstruction is developed. This is extended to the reconstruction of multiple layers and whole structures. Different algorithms are used to determine all materials layer by layer or at once. With the basic knowledge of the specimens structure, parameter optimization algorithms are used to obtain maximal correlation between measurement and simulation. Finally, the tree-dimensional structure of a complete specimen is obtain with a classification approach.

Several applications for the new analysis system are shown in chapter 6. The results have been obtained with the newly developed ultrasonic analysis program “USAnalyser”, which is based on the algorithms and theoretical considerations of this thesis. Different analysis modes are compared with respect to their precision and information content. The analysis of stacked components with molding compound is demonstrated, which delivers information on flaws inside. The precise time-of-flight measurements are used for new applications, like material characterization and homogeneity analysis. Based on the additional evaluation of the focus positions, the new method for simultaneous determination of sound velocity and thickness is demonstrated.

The conclusions of this thesis are presented in chapter 7. Some possible future developments and perspectives are discussed, which could help to improve ultrasonic analysis even more.

2 Sound field

2.1 Sound field measurement

Sound field simulations, which are not based on model assumptions of the transducer, require a sound field measurement as input. Additionally, all required sound field parameters for (partially) transducer model based ultrasonic simulations can be obtained.

There are several ways to measure the sound pressure field. One commonly used technique is to measure the sound field not directly, but the received sound field convoluted with the impulse response of the same or another transducer.

Based on the linear system approach, unknown and known components in equation (1) can be separated. In most cases only the received signal $s_r(t)$ can be measured and the excitation signal *I*, the transducer impulse response *TD* and the receiver circuit impulse response *RE* are unknown. Hence there is reversibility: *I, TD, SP* and *RE* can be put at any position in equation (1). Now all unknown components can be combined into one sound field describing impulse response matrix *SF*:

$$s_r(t) = \int_{x_1}\int_{y_1}\int_{x_2}\int_{y_2} SF_1(t,x_1,y_1) * SP(t,x_1,y_1,x_2,y_2) * SF_2(t,x_2,y_2) dy_2 dx_2 dy_1 dx_1 \, . \qquad (2)$$

On an one transducer system, *SF* is commonly very close to the actual sound pressure field that a transducer emits, since *I* and *RE* are very broad band (similar to a time shifted Dirac impulse). With this assumption, the sound pressure field can be determined from the self-deconvolution of *SF*SF* as long as $s_r(t)$ and *SP* are known. The received signal $s_r(t)$ can be measured, and a well known specimen *SP* is a thin wire with a ball head smaller than the wave length in water. It has an impulse response that is zero everywhere except at the position of the ball head, where the sound is reflected. The self deconvolution of the reflection signal represents the time pressure function of the sound field at the position of the ball head (figure 4). It is sufficient to know the sound pressure field at one *z*-level. The complete sound field can be determined with a sound field simulation. Only one single line scan through the center axis is sufficient, if the transducer has rotatory symmetry (figure 5).

Since the measurement represents the self convoluted sound field *SF*SF*, the self deconvolution has to be applied. One possible approach is the determination of the square root form the Fourier transform:

$$SF(x,y) = \mathrm{FFT}^{-1}\sqrt{\mathrm{FFT}\left[s_r(x,y)/SP_{\max}\right]}, \text{ with } SP_{\max} = \text{reflectivity of the ball head.} \qquad (3)$$

If s_r consists of N samples, equation (3) provides 2^N solutions. Since all natural pulse signals have a continuous spectrum, we can apply the minimal phase criterion to reduce the number of solutions. The two remaining solutions are equal but with inverted amplitudes. All signals of the sound field are matched by cross-correlation to the same solution (figure 8).

The most important sound field characteristics are the shape of the wavefront (figure 9) and the amplitude distribution in dependence propagation direction (figure 10). They can be determined directly from the measured signals without self-deconvolution of the signals. For this, the times have to be divided by two and the amplitudes are the square root of the measured amplitudes. The shape of the wave front is determined from the time-of-flight of the peak of the signals and by multiplying with the sound velocity of water. The amplitude A_0 of the amplitude distribution in dependence of the angle θ_0 from the shape $r(\theta_0)$ is obtained from the peak amplitude A and the energetic area weight:

$$A_0(\theta_0) = \frac{A(\theta_0) \cdot \pi \cdot \sqrt{2\pi \cdot r(\theta_0)}}{2 \cdot \int_0^{\pi/2} A(\theta) \cdot \sqrt{2\pi \cdot r(\theta)} \cdot d\theta}. \qquad (4)$$

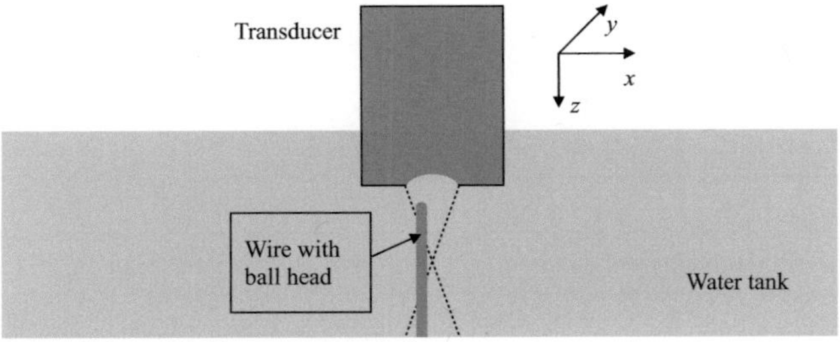

Figure 4: Sound field measurement scheme

The self-convoluted sound-field $SF*SF$ can be measured, with the principle shown in figure 4. The transducers used all have rotatory symmetry. Hence, a single line measurement through the center is enough to determine the complete sound-field. Theoretically, the z-level for the measurement can be at any height for this. For practical reasons, like signal-to-noise ratio (*SNR*), positions close to the transducer are providing the best results.

The signals $s_r(t)$ from such a measurement are shown in figure 5. It is created from a point by point line measurement at a constant z-level, averaging 1000 times per point to improve the *SNR*. The measurement clearly shows the sound field with some random noise and horizontal lines caused by the remnants of systematic noise. In this case the main part of the coherent systematic noise is already subtracted from the original data but the remnants would produce strong artifacts in a sound field simulation. The random noise is less problematic due to its averaging to zero behavior. For reduction of noise, a geometrical low-pass filter was applied to the data, which blocks everything but the desired spherical shaped wave front. The main noise remnants have random behavior (figure 6).

Now the self-deconvolution is applied, which divides all times by two causing a shape change (figure 7). It requires a good *SNR*, otherwise the minimal phase criterion can not be fulfilled and the signals are distorted. This happens first at the edges of the sound field. If this data is used for an ultrasonic simulation, strong artifacts would be generated at the edges. This problem is reduced by applying a wave front shape-matched geometrically low-pass filter to the data one more time, which reduces all contents with a different shape (figure 8). This data is used for all sound field simulations in this thesis.

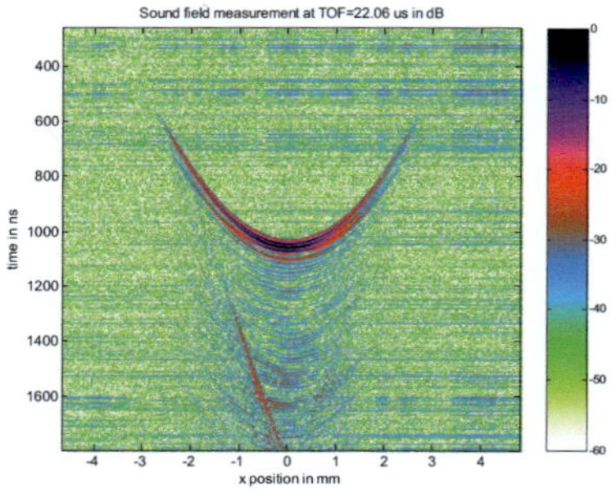

Figure 5: Measurement of the self-convoluted sound field; 75 MHz transducer; scale in dB; 1000 times averaged data

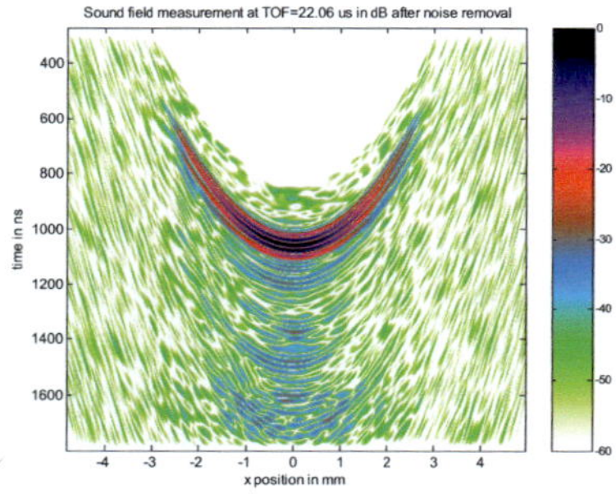

Figure 6: Self-convoluted sound field after noise filtering; 75 MHz transducer; logarithmic scale in dB

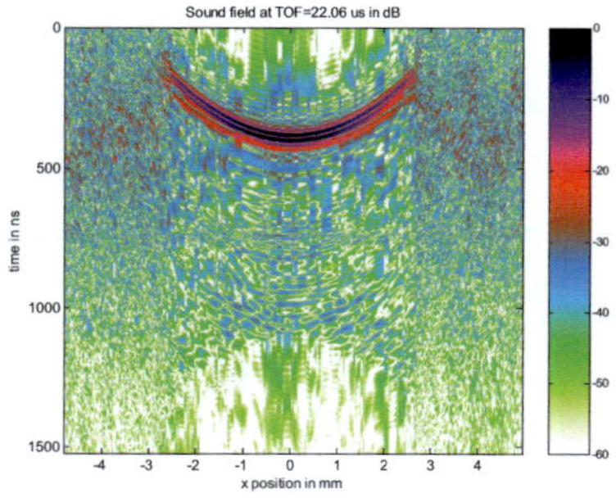

Figure 7: Sound field (self-deconvolution of measurement); 75 MHz transducer; logarithmic scale in dB

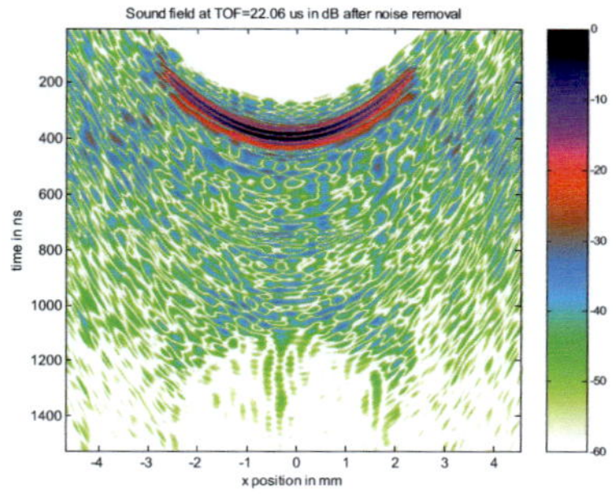

Figure 8: Sound field at a constant *z*-level after noise filtering; 75 MHz transducer; logarithmic scale in dB

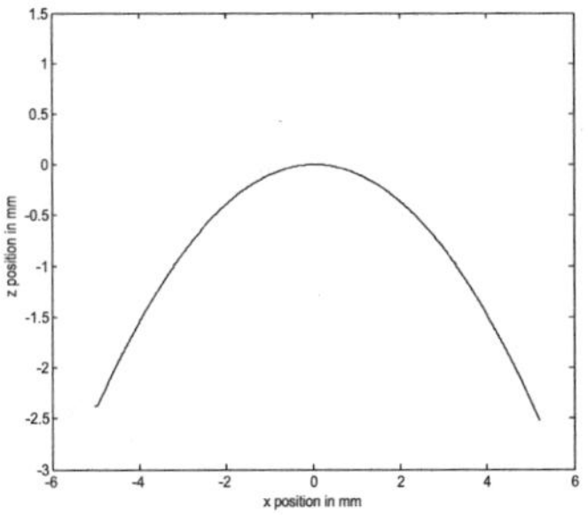

Figure 9: Wave front shape; 75 MHz transducer analytic approximation with a parabola from three near-center points from cross-correlation

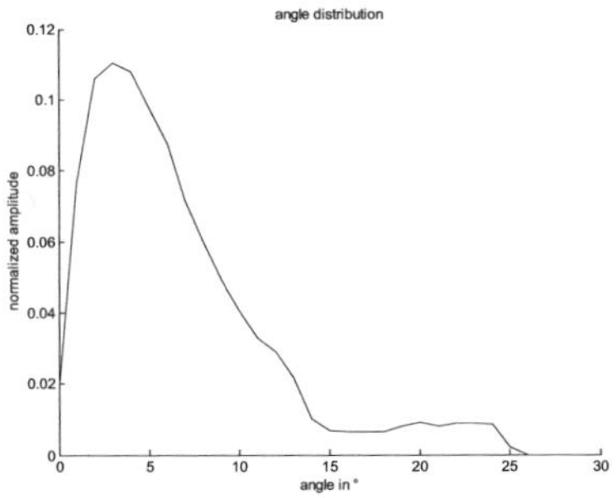

Figure 10: Amplitude distribution in dependence of the propagation direction from equation (4); 75 MHz transducer

2.2 Sound field modeling

Elastic wave propagation can be described with the linear homogeneous elastodynamic wave equation for the displacement u:

$$\rho \frac{\partial^2 u_j}{\partial t^2} - \frac{\partial}{\partial x_i}\left(C_{ijkl} \frac{\partial u_k}{\partial x_l} \right) = 0, \text{ for } i, j, k, l = 1, 2, 3. \tag{5}$$

Here ρ is the mass density, t the time, x the coordinates and C the stiffness tensor. It can be derived from the equation of motion and the linear stress and strain relations of Hooke's law. With the solutions of the wave equation (5) for the different wave modes and boundary conditions, the wave propagation can be modeled inside different materials and at material interfaces.

2.2.1 Reflection and transmission coefficients

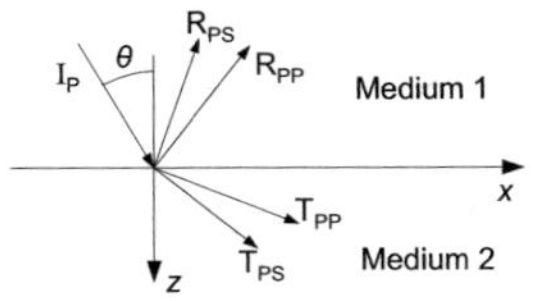

Figure 11: Reflection and transmission for pressure-wave incidence

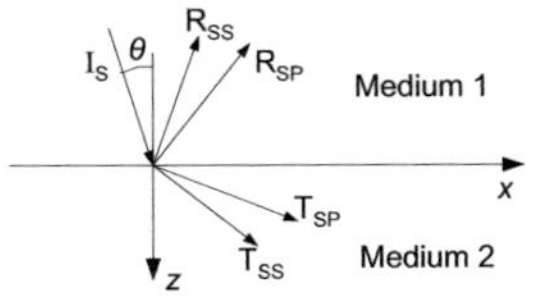

Figure 12: Reflection and transmission for shear-wave incidence

The reflection and transmission coefficients are required to describe the interaction of an ultrasonic wave with a material interface. For plane waves, they are determined at plane interfaces (figure 11 and 12), from the boundary conditions at the interface and the wave equation in the material. Due to mode conversion and reflection, an incoming pressure wave I_P or shear wave I_S splits up into four waves: reflected shear wave (R_{PS} or R_{SS}), reflected pressure wave (R_{PP} or R_{SP}), transmitted shear wave (T_{PS} or T_{SS}) and transmitted pressure wave (T_{PP} or T_{SP}). The amplitudes of the transmitted (T) and reflected (R) waves are determined with the reflection (*R*) and transmission (*T*) coefficients from the amplitude of the initial wave. The wave normal can be tilted in the *x*-*z*-plane and the wave is independent of the *y*-axis.

A detailed derivation of the reflection and transmission coefficients can be found in the appendix on the pages 90-94. It is based on the wave equation (5), Hooke's law (75) for isotropic materials and the boundary conditions (74) at the material interface. From those equations, the equation systems (88) and (93) are derived, and the unknown variables of them are the reflection and the transmission coefficients for the sound particle velocity. The solutions are listed below in the equations (6):

$$R_{PP} = \frac{1}{N}\left(-I_1 - I_2 + I_3 + I_4\right) \tag{6a}$$

$$R_{PS} = \frac{1}{N}\left(2c_{P1}\xi w_{P1} I_5\right) \tag{6b}$$

$$T_{PP} = \frac{2c_{P1}\rho_1 w_{P1}\cdot\left(2\,c_{S1}^2\xi^2 + q_{S1}\right)\left[\left(q_{S2}\rho_2 + 2\,c_{S1}^2\xi^2\rho_1\right)c_{S2}w_{S1} + \left(q_{S1}\rho_1 + 2c_{S2}^2\xi^2\rho_2\right)c_{S1}w_{S2}\right]}{N} \tag{6c}$$

$$T_{PS} = \frac{1}{N}\left(2c_{P1}\rho_1 w_{P1}\xi\left(2\,c_{S1}^2\xi^2 + q_{S1}\right)\left[\left(-q_{S1}\rho_1 + q_{S2}\rho_2\right)c_{P2}c_{S1} + \left(c_{S1}^2\rho_1 - c_{S2}^2\rho_2\right)2w_{P2}w_{S1}\right]\right) \tag{6d}$$

$$R_{SP} = \frac{1}{N}\left(2c_{S1}\xi w_{S1} I_5\right) \tag{6e}$$

$$R_{SS} = \frac{1}{N}\left(I_1 - I_2 + I_3 - I_4\right) \tag{6f}$$

$$T_{SP} = \frac{1}{N}\left(2c_{S1}\rho_1 w_{S1}\xi\left(2\,c_{S1}^2\xi^2 + q_{S1}\right)\left[\left(q_{S1}\rho_1 - q_{S2}\rho_2\right)c_{P1}c_{S2} + \left(-c_{S1}^2\rho_1 + c_{S2}^2\rho_2\right)2w_{P1}w_{S2}\right]\right) \tag{6g}$$

$$T_{SS} = \frac{2c_{S1}\rho_1 w_{S1}\cdot\left(2\,c_{S1}^2\xi^2 + q_{S1}\right)\left[\left(q_{S2}\rho_2 + 2c_{S1}^2\xi^2\rho_1\right)c_{P2}w_{P1} + \left(q_{S1}\rho_1 + 2c_{S2}^2\xi^2\rho_2\right)c_{P1}w_{P2}\right]}{N} \tag{6h}$$

with

$$N = I_1 + I_2 + I_3 + I_4$$
$$I_1 = c_{P1}c_{P2}c_{S1}c_{S2}\xi^2 \cdot (q_{S1}\rho_1 - q_{S2}\rho_2)^2$$
$$I_2 = c_{P1} \cdot \left[c_{S2}\rho_1\rho_2 w_{P2}w_{S1} \cdot (2c_{S1}^2\xi^2 + q_{S1})(2c_{S2}^2\xi^2 + q_{S2}) + c_{S1}w_{P2}w_{S2} \cdot (q_{S1}\rho_1 + 2c_{S2}^2\xi^2\rho_2)^2 \right]$$
$$I_3 = w_{P1}w_{S2}\rho_2 c_{P2}\rho_1 c_{S1} \cdot (2\,c_{S1}^2\xi^2 + q_{S1})(2c_{S2}^2\xi^2 + q_{S2})$$
$$I_4 = w_{P1}w_{S1} \cdot \left[c_{P2}c_{S2} \cdot (2c_{S1}^2\xi^2\rho_1 + q_{S2}\rho_2)^2 + 4w_{P2}w_{S2}\xi^2 \cdot (c_{S1}^2\rho_1 - c_{S2}^2\rho_2)^2 \right]$$
$$I_5 = c_{P2}c_{S2} \cdot (q_{S1}\rho_1 - q_{S2}\rho_2)(2c_{S1}^2\xi^2\rho_1 + q_{S2}\rho_2) + 2w_{P2}w_{S2} \cdot (c_{S1}^2\rho_1 - c_{S2}^2\rho_2)(q_{S1}\rho_1 + 2c_{S2}^2\xi^2\rho_2)$$

$q_{ab} = (1 - 2c_{ab}^2\xi^2)$, $w_{ab} = \sqrt{1 - c_{ab}^2\xi^2}$, $\xi = \dfrac{\sin(\theta_{ab})}{c_{ab}}$. where "ab" can be P1, P2, S1 or S2.

For one-dimensional plane waves with orthogonal incidence ($\theta_0 = 0$) the reflection and transmission coefficients for the sound particle velocity simplify to:

$$R_{PP} = \frac{-c_{P1}\rho_1 + c_{P2}\rho_2}{c_{P1}\rho_1 + c_{P2}\rho_2}, \tag{7a}$$

$$R_{PS} = 0, \tag{7b}$$

$$T_{PP} = \frac{2c_{P1}\rho_1}{c_{P1}\rho_1 + c_{P2}\rho_2}, \tag{7c}$$

$$T_{PS} = 0. \tag{7d}$$

Since all mode conversion coefficients are zero, no shear waves are generated from plane pressure waves with $\theta = 0$.

An example for the determination of the amplitudes of the reflected waves of a glass slide in dependence of the initial angle can be found in figure 25.

2.2.2 Sound field modeling with plane waves

Each sound field can be modeled with the superposition of point source fields (e.g. [2], [3]). For that, the distance between the single points should be about one tenth of the shortest wavelength, to achieve accurate simulation results. The number of points required for that is relatively high, what leads to high mathematical effort. According to reference [3], the sound propagation can also be described with Gaussian wave packets, which are based on plane waves. It is a far field approach and can be used for the decomposition of the sound field into plane waves. Those plane waves are governed with an envelope function, and the original sound field is modeled by the superposition of all wave packets. In the case of smooth wave shapes, as with the spherical focused transducers used here, only a low number of wave packets are required, which greatly reduces the mathematical effort.

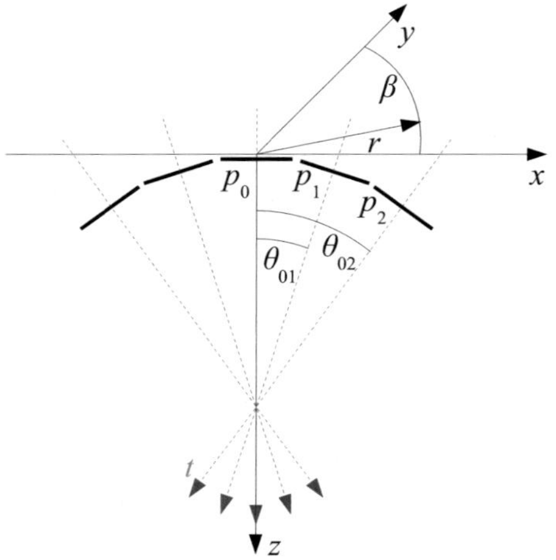

Figure 13: Sound field decomposition scheme

In this thesis, the decomposition formulas are derived for the scalar sound pressure field p into plane waves p_i with the packet index i as illustrated in figure 13. The equations can easily be extended to vector fields, since plane waves have a constant propagation direction and constant oscillation direction. The sound field pressure is now:

$$p(t,x,y,z) = \sum p_i(t,x,y,z) \text{ with} \tag{8}$$

$$p_i(t,x,y,z) = \text{env}_i(x,y,z) \cdot s_i\left(t - \frac{l_{xi}\cdot(x-x_{0i}) + l_{yi}\cdot(y-y_{0i}) + l_{zi}\cdot(z-z_{0i})}{c\sqrt{l_{xi}^2 + l_{yi}^2 + l_{zi}^2}}\right), \tag{9}$$

where (l_{xi}, l_{yi}, l_{zi}) represents the wave normal, (x_{0i}, y_{0i}, z_{0i}) the initial position, s_i the time pressure signal of the plane wave and t the time. As envelope function env_i the Gaussian distribution is very convenient, because it does not cause significant diffraction on wave propagation and the reflection and transmission coefficients of plane waves remain basically valid [3]:

$$\text{env}_i(x,y,z) = \exp\left(-\frac{(x-x_{0i}-hl_{xi})^2 + (y-y_{0i}-hl_{yi})^2 + (z-z_{0i}-hl_{zi})^2}{\sigma^2}\right), \tag{10}$$

with $h = \dfrac{l_{xi}(x-x_{0i}) + l_{yi}(y-y_{0i}) + l_{zi}(z-z_{0i})}{l_{xi}^2 + l_{yi}^2 + l_{zi}^2}$ and σ the window width.

Since spherical transducers are commonly used for ultrasonic microscopes, rotatory symmetry can be assumed. This transforms the coordinates as follows: $\theta = \arctan(r / z)$, $\beta = \arctan(y / x)$ and $r^2 = x^2 + y^2$. With this assumption, p_i can be described with the superposition of infinite uniform distributed plane waves with the same (r, z, θ) and different $\beta = \arctan(l_{yi} / l_{xi})$:

$$p_i(t,r,z) = \frac{1}{2\pi} \int_{\beta=0}^{2\pi} \mathrm{env}_i(r,z,\beta) \cdot s_i\left(t - \frac{\sin(\theta_{0i})(r \cdot \cos(\beta) - r_{0i}) + \cos(\theta_{0i})(z - z_{0i})}{c}\right) d\beta \,. \quad (11)$$

with $\mathrm{env}_i(r,z,\beta) = e^{\frac{[(r-r_{0i})\cos(\theta_{0i})-(z-z_{0i})\sin(\theta_{0i})]^2+(1-\cos(\beta))\{\sin^2(\theta_{0i})[r^2\cdot(1+\cos(\beta))-2rr_{0i}]+2r\cdot[\cos(\theta_{0i})\sin(\theta_{0i})(z-z_{0i})+r_{0i}]\}}{\sigma^2}}$,

(r_{0i}, z_{0i}) the initial position and $\theta_{0i} = \arctan(l_{ri} / l_{zi})$ the sound wave angle.

The field equation for a single plane wave is obtained with a constant β (e.g. $\beta = 0$):

$$p_i(t,r,z) = \exp\left(-\frac{[(r-r_{0i})\cos\theta_{0i} - (z-z_{0i})\sin\theta_{0i}]^2}{\sigma^2}\right) \cdot s_i\left(t - \frac{(r-r_{0i})\sin\theta_{0i} + (z-z_{0i})\cos\theta_{0i}}{c}\right). \quad (12)$$

Since all other plane waves with a different β have the same characteristics as the plane wave with $\beta = 0$, equation (12) can be used for the determination of wave propagation characteristics. The integral of equation (11) for the rotatory symmetric solution can not be solved analytically, but it can be approximated numerically with a reduced interval for β. Only plane waves with a distance $|\beta r|$ closer than σ to the observation point (r, z) have a significant effect on the result. This behavior can be used to reduce the mathematical effort.

For verification of the practical usability of the decomposition algorithm, comparative simulations have been done. In figure 14 the results are compared with the results of the point source synthesis. The point spacing is 1 µm and the calculation of the transfer functions for an infinite media is done according to Huygen's principle. The initial field data for both simulations is taken from the same sound field measurement (figure 8). For the plane wave decomposition, the window width $\sigma = 0.5$ mm for the envelope function has been chosen large enough, to ensure proper superposition of the single wave packets with an angle spacing of 1°. The signal s_i and the angle θ_i are obtained at the center of each wave packet from the measured field data.

For the chosen three z positions, the data of the point synthesis and of the plane wave decomposition (figure 14) has been compared with a two-dimensional cross-correlation. There are a lot of visible differences, but the major wave front is highly correlated. The allover correlation is about 90%. Most of the differences are caused by artifacts. In the point synthesis results are strong diffraction effects at the edges of the sound field caused by an insufficient sound field measurement. The signal-to-noise ratio at the edges is not sufficient and this causes an abrupt end of the measured field (figure 8), where actual sound field continues. Those effects can not occur in the plane wave simulation, since there is no diffraction (far field approximation). Another type of artifacts is visible in the center of the point synthesis simulation results. It is caused by the sound field measurement random noise that concentrates in the center and does not average to zero because of the cylindrical symmetric system. This effect is also visible in the plane wave simulation, but here the initial signal has been chosen shorter, whereby it does not cause so strong effects.

The plane wave simulation has also been compared to a finite element (EFIT [4]) simulation for three chosen times (figure 15). The EFIT simulation shows the same effects like the point source synthesis, because it is initialized with the same sound field measurement data. An additional artificial effect is numerical dispersion, which causes "ringing" effects that increase with simulation time.

For the given sound field, all simulation approaches provide usable results, but with different side effects and simulation time. The mathematical effort for the plane wave simulation is about 1000 times lower than for the point source synthesis. The full numerical EFIT simulation has a much higher mathematical effort than the methods mentioned above. It is only used for comparison and verification of results.

If only the received signal and no sound field is required, the effort for the sound field decomposition into plane waves can be reduced to a ray trace based impulse response determination (chapter 3.3.2) and the convolution with reference signals (chapter 3.2.2).

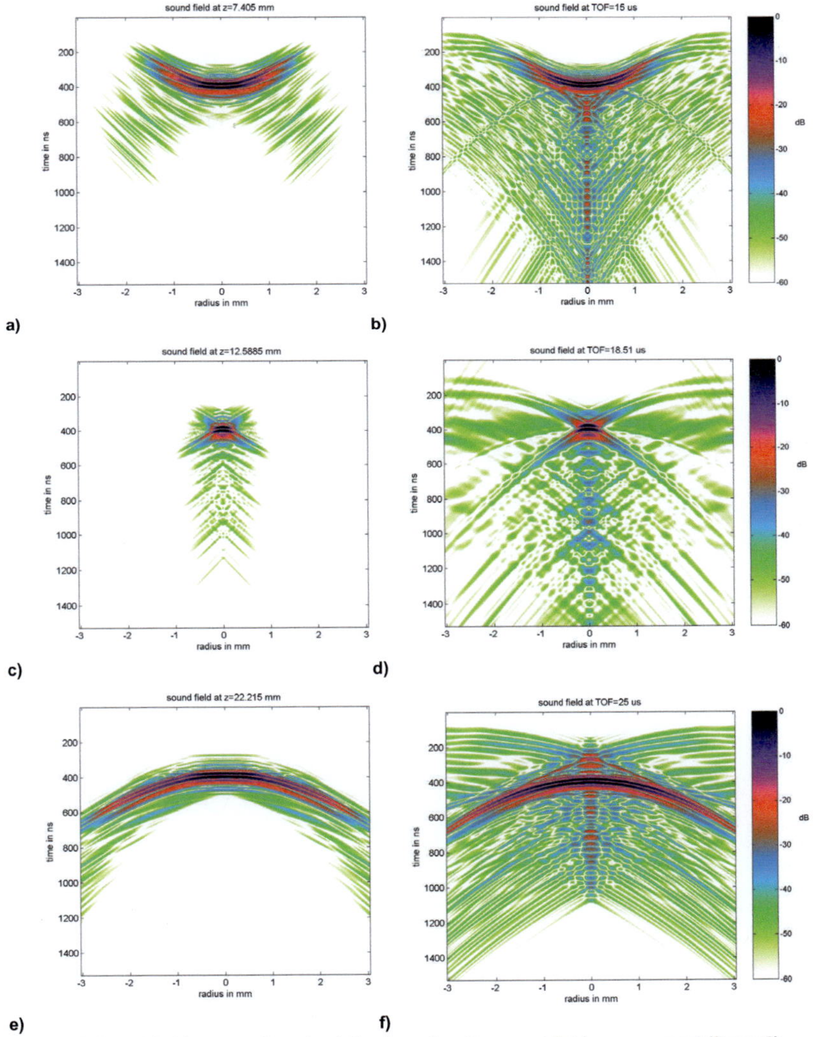

Figure 14: Sound field propagation simulation based on the sound field measurement (figure 8)
a),c),e) Sound field decomposition into plane waves with equation (11), σ = 0.5 mm, dθ = 1°
b),d),f) Point source synthesis (Huygens principle, point distance 1 µm)
Correlation of the field data at the same *z* level: a) & b) = 0.92, c) & d) = 0.89, e) & f) = 0.81

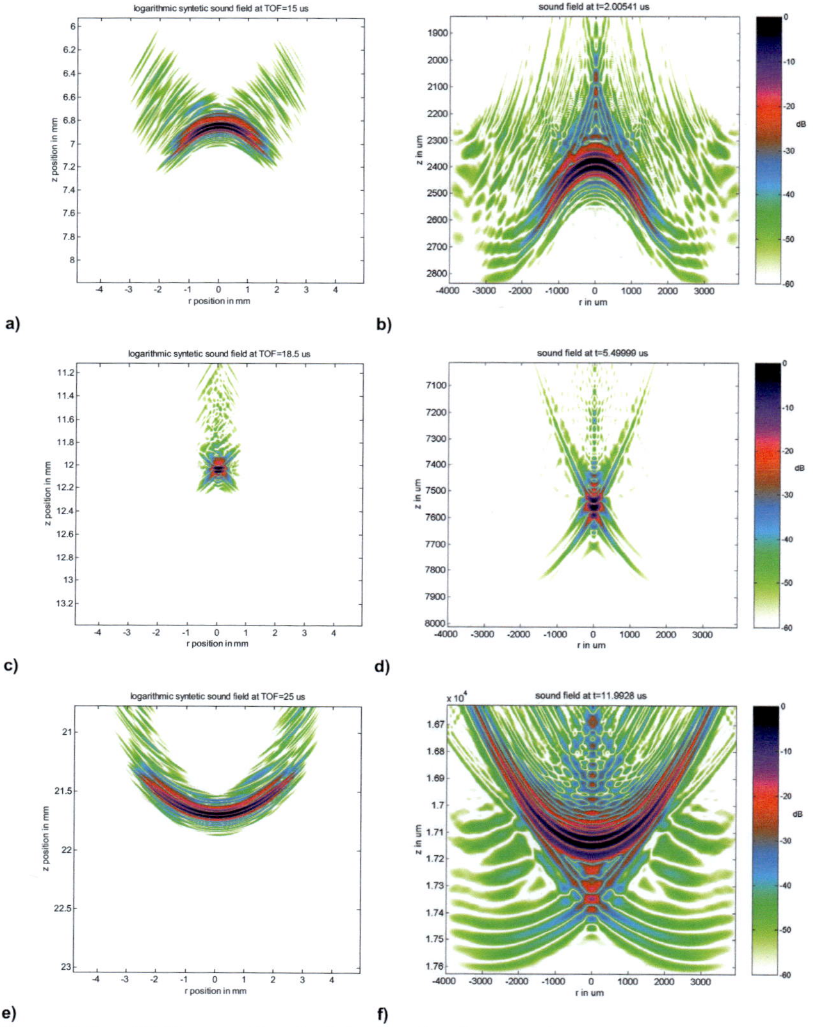

Figure 15: Sound field propagation simulation based on the sound field measurement (figure 8)
a),c),e) Sound field decomposition into plane waves with equation (11), σ = 0.5 mm, dθ = 1°
b),d),f) C-EFIT [4] simulation snap shots (*z* offset -4.5 mm, time offset -13.0 µs, different scaling)

2.2.3 Generalized sound field position

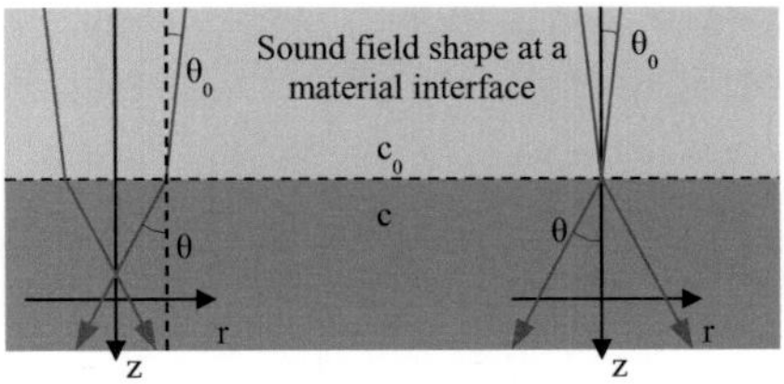

Figure 16: Idealized sound field description with ray geometry at a planar material interface with different focus positions

The prediction of the sound field shape inside different materials and structures is essential for ultrasonic simulation and reconstruction approaches. An exact representation is not always necessary or useful, so simplifications can be used to obtain an efficient modeling of the sound field shape.

The sound field of spherical focused transducers is generally very wide compared with the wavelengths of an ultrasonic pulse (figure 8). With this assumption, the sound field can be described with multiple rays representing the single plane wave packets of the sound field decomposition (chapter 2.2.2). One ray, as in figure 16, represents a component of the complete sound field with the initial angle of incidence θ_0. The radial position r of each ray as a function of time t is the integral of the sound velocity c of the sound path over the time-of-flight (t - t_0):

$$r(t) = r(t_0) - \int_{t_0}^{t} c(\tau) \cdot \sin\left(\theta(\tau)\right) d\tau \,. \tag{13}$$

where t_0 is the start time, θ the angle of incidence, $\theta_0 = \theta(t_0)$ the start angle and $r(t_0)$ the start position. According to Snell's law, $\sin(\theta)$ can be replaced by the constant initial condition of each ray:

$$r(t) = r(t_0) - \frac{\sin\left(\theta(t_0)\right)}{c(t_0)} \int_{t_0}^{t} c(\tau)^2 \, d\tau \tag{14}$$

Now, the initial condition can be separated from the time dependent part and the generalized sound field position GSP can be defined:

$$GSP(t) = \int_{0}^{t} c(\tau)^2 \, d\tau \,. \tag{15}$$

The GSP is a sound field position value, which provides the sound field shape for any structure and any sound path. This means, that the sound field has always the same lateral shape for the same *GSP* values. For the simple case of only one plane wave with one initial angle, the sound field is always identical for the same values of the *GSP*. Under real circumstances (angle distribution) this is only true on average, since $c(\tau)$ also depends on $\theta(t_0)$. The *GSP* provides the ability to transform a known sound field shape into a new shape, which it would have in a different material structure.

In a continuous media the integral simplifies to $GSP(t) = c^2 t$. As an example, this equation can be used to express a material and structure independent generalized position of the natural focus GSP_N. With the well known formula for the natural focus distance (near field length) N of a plane circular transducer with the diameter D and at the frequency f, GSP_N is:

$$GSP_N = c \cdot N = D^2 \cdot f / 4 . \tag{16}$$

It only depends on the transducer parameters D and f, and is independent of the used material and structure.

With the *GSP* it is possible to predict the focus position in different material setups if the focus position is known in water. More generally it is possible to estimate the time-of-flight for an equivalent surface reflection (later called reference signal) with similar properties like an arbitrary interface reflection. In this case the *GSP* must be the same for the reference signal and for the interface echo.

Some practical correlation results of different signals are shown in the figures 29 and 30. Especially for the transducers with a long delay line (like the 75 MHz transducer), the ray propagation assumptions are mostly valid and different signals with the same *GSP* value are highly correlated. The major differences are caused by frequency dependent ultrasonic attenuation in the materials and the sound velocity $c(\tau)$ dependencies from the initial sound field angle.

2.3 Receiving transducer signal

The major characteristics of the signal received from a single reflection are the amplitude and the signal shape. They can be determined from the sound field or they are obtained directly from reference signal measurements.

2.3.1 Calculation of the transducer signal from the sound field

The received signal s_r of the transducer is determined with the summation of the orthogonal tension at the transducer surface, which corresponds to the sound pressure field p in water. It is convoluted with the impulses response *SF* of the transducer, which includes the conversion factor between pressure and voltage, the electrical system and the delay line and/or lens. For a cylindrically symmetric transducer at the position $(z,0)$ and with the radius R, the signal is:

$$s_r(t,z) = \int_{r=0}^{R} SF(t,r) * \frac{2\pi r}{R} \cdot p(t,r,z) dr \,. \tag{17}$$

Based on the sound field decomposition into governed plane waves, the received signal can be obtained from the summation over all initial angle θ_{0i} indices i of the single signals, which governed plane waves p_i from equation (11) generate:

$$s_r(t,z) = \sum_i s_{ri}(t,z) = \sum_i \int_{r=0}^{R} SF(t,r) * \frac{2\pi r}{R} \cdot p_i(t,r,z) dr \,. \tag{18}$$

2.3.2 Received signal amplitude

The amplitude of the received signal mainly depends on the constructive interference of the sound field at the transducer surface. It changes with different angles of the wave normal. This dependency is required for the ultrasonic simulation to achieve a correct superposition of the received waves.

We can determine the maximum amplitude in dependence of the angle θ of the wave normal, if we assume a Gaussian pulse like signal $s_i(t)$:

$$s_i(t) = A_{0i} \cdot \cos\left(2\pi f_c \cdot (t-t_0) + \varphi_0\right) \cdot \exp\left(-\frac{(t-t_0)^2}{\tau^2}\right), \tag{19}$$

where f_c is the center frequency, τ the pulse width, φ_0 the initial phase and A_{0i} the amplitude. The relative maximum of s_{ri} is obtained at $t = t_0$, with $\varphi_0 = 0$ ($z = z_0$ for simplification) from equation (18) and (12):

$$s_{ri\_\max}(\theta_i) \sim \int_{r=0}^{R} 2\pi A_{0i} r \cdot \exp\left(-r^2 \frac{c^2\tau^2 + \sin^2(\theta_i) \cdot \left(\sigma^2 - c^2\tau^2\right)}{c^2\tau^2\sigma^2}\right) \cdot \cos\left(2\pi f_c r \cdot \sin(\theta_i)\right) dr \,. \tag{20}$$

This integral has no direct analytic solution, since the result involves the error function:

$$s_{ri\_\max}(\theta_i) \sim \frac{\pi A_{0i}}{k^2}\Bigg[1 - \exp\left(-k^2R^2\right) \cdot \cos\left(2\pi f_c R \cdot \sin(\theta_i)\right) + \frac{i\pi f_c \sqrt{\pi}}{k} \cdot \sin(\theta_i) \cdot \exp\left(-\frac{\pi^2 f_c^2 \cdot \sin^2(\theta_i)}{k^2}\right) \cdot \mathrm{Im}\left(\mathrm{erf}\left(\frac{i\pi f_c \cdot \sin(\theta_i)}{k}\right) - \mathrm{erf}\left(\frac{i\pi f_c \cdot \sin(\theta_i)}{k} - kR\right)\right)\Bigg], \tag{21}$$

with $k^2 = \dfrac{c^2\tau^2 + \sin^2(\theta_i)\cdot(\sigma^2 - c^2\tau^2)}{c^2\tau^2\sigma^2}$.

Under the assumption of small waves against the transducer size ($R \to \infty$), equation (21) simplifies to:

$$s_{ri\_\max}(\theta_i) \sim \frac{\pi\sigma^2 A_{0i}}{1+\left(\dfrac{\sigma^2}{c^2\tau^2}-1\right)\cdot\sin^2(\theta_i)}. \tag{22}$$

According to figure 17, the angle θ of the received wave normal can be determined from the distance b between specimen and spherical focused transducer in relation to the curve radius g:

$$\theta = \alpha - \theta_0 = \arcsin\left(2\cdot\frac{b-g}{g}\cdot\sin(\theta_0)\right). \tag{23}$$

For a more general solution we can replace the distance by the material structure compensating *GSP* value for the sound field and compare it to the GSP_F value at the focus plane (focus point):

$$s_{ri\_\max}(GSP) \sim \frac{\pi\sigma^2 A_{0i}}{1+\left(\dfrac{\sigma^2}{c^2\tau^2}-1\right)\left(2\dfrac{GSP-GSP_F}{GSP_F}\cdot\sin(\theta_{0i})\right)^2}, \tag{24}$$

with $GSP_F = \int_{z=0}^{g} c(z)dz$ and $GSP = \int_{z=0}^{b} c(z)dz$.

Equation (24) provides the relative amplitude factor for each received wave. Most of the constants are unknown and have to be determined from a reference measurement. For the normalized amplitude factor A_{ni}, the unknowns can be combined to k_A and GSP_F:

$$A_{ni}(GSP) = \frac{1}{1+k_A\cdot\left[(GSP-GSP_F)\cdot\sin(\theta_{0i})\right]^2}, \text{ with } k_A = \frac{4}{GSP_F^{\,2}}\left(\frac{\sigma^2}{c^2\tau^2}-1\right). \tag{25}$$

The factors k_A and GSP_F are specific for the transducer used and the excitation pulse. They are independent of the propagation media (e.g. specimen). Once they have been determined, they can be used for all simulations. A comparison between a measured amplitude curve and a simulated one is shown in figure 18. There is a high conformance in general, except for the small aberration in the measurement caused by the phase inversion at the focus point. The precision mainly depends on the accuracy of the measured amplitude distribution of the sound field (figure 10).

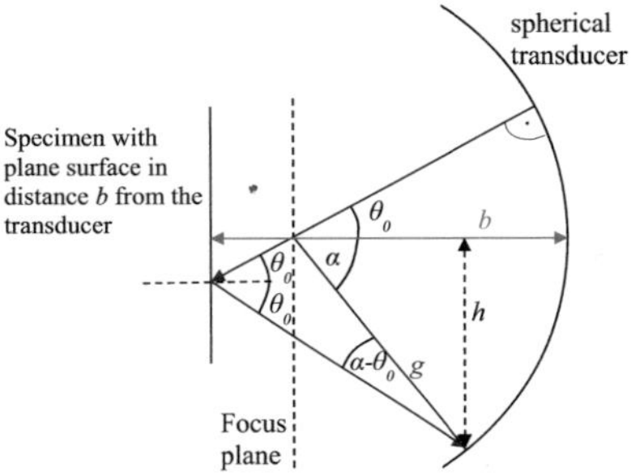

Figure 17: Geometrical considerations at a spherical transducer

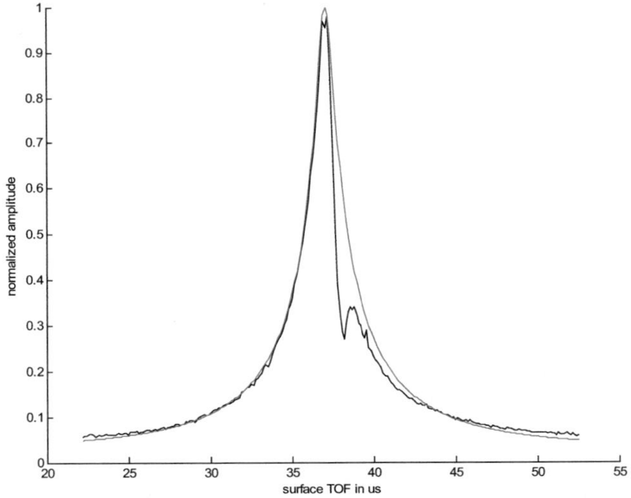

Figure 18: Amplitude curve – maximum amplitude in dependence of the *TOF* (transducer *z*-position) black: measurement of a glass surface with the 75 MHz transducer and attenuation compensation; grey: 2D ray-trace simulated amplitude curve of a glass surface with equation (25) based on the sound field measurement (figure 10), 25 initial ray's in 1° steps

2.3.3 Measurement of reference signals

A reference signal represents the reflection signal of a plane surface interface under orthogonal incidence. They are measured with a *z*-scan in the possible distance range between transducer and plane surface. The reference signals consist of the interference of the sound field at the transducer surface and define the signal shape for different sound field positions. The assumption of plane interfaces is restrictive, but very convenient for many applications. Since reference signals are recorded directly from a strong surface echo, they usually have a very good signal-to-noise ratio compared to sound field measurements with a very small point reflector. The signal shape is specific for each sound field position and thereby for each *GSP* value. Each reference signal has a different time-of-flight in water t_{total} and a different GSP_R value witch is obtained with temperature matched sound velocity in water c_w:

$$GSP_R(c_w, t_{total}) = t_{total} \cdot c_w^{\ 2} \tag{26}$$

To cover all possible *GSP* values, representative references are stored in a reference database. *GSP* values that fall in-between two representative references can be interpolated, while the data has to be extrapolated for *GSP* values beyond the representative references. It has been observed that linear interpolation between two reference signals with time shifted components can lead to blurred signals. This effect can be avoided with spectral interpolation and the minimal phase criterion, since all natural pulse signals have continuous spectra.

The material of the reference specimen can be of any kind (e.g. glass). Under some circumstances (e.g. strong surface waves) it might be more convenient to choose a material with material parameters similar to those of the investigated specimen. Surface and interface wave modes are not considered yet in this thesis, although they do have some small effect on the reflection signal (slanted lines in figure 28).

For an optimal determination of reference signals, the measured reflection signals have to be processed to remove most of the noise and distortions. A digital reflection signal $s_r(n)$ from the ultrasonic microscope consists of the following components:

$$s_r(n) = rs(nT - \Delta t - \Delta t_F) + pr(nT - \Delta t) + sn(nT - \Delta t) + rn + qn\,, \tag{27}$$

where n is the sample number, T the sampling time, Δt the random sub-sample time delay (jitter), Δt_F the time-of-flight fluctuations of the sound path in water, $rs(t)$ the desired reflection signal, $pr(t)$ the permanent internal transducer echoes, $sn(t)$ the time coherent noise from the electrical system, rn the random noise and qn the quantization noise. The following work steps are performed to obtain $rs(n)$ from $s_r(n)$:

1. Measurement of N reflection signals $s_r(n)_1 .. s_r(n)_N$ under the same condition.
2. Sub-sample time delay correction between $s_r(n)_1 .. s_r(n)_N$ and the permanent signal $pr(n)$ with cross-correlation time delay estimation and sub-sample time delay shift. $pr(n)$ is determined from an empty reference measurement (no specimen present) and contains $pr(t)$ and $sn(t)$.
3. Subtraction of $pr(n)$ from the reflection signals $s_r(n)_1 .. s_r(n)_N$.
4. Sub-sample time delay correction between $s_r(n)_2 .. s_r(n)_N$ and $s_r(n)_1$ to correct time-of-flight fluctuations $_\Delta t_F$ in the water delay line.
5. Averaging of N signals $s_r(n)_1 .. s_r(n)_N$ to reduce random noise from rn and qn: $rs(n) = \Sigma\, s_r(n)_i \,/\, N$.
6. Update $pr(n)$ to correct system changes due to transducer movement.

The signal $rs(GSP_R, t)$ represents now the reference signal in dependence of the generalized sound field position GSP_R. All actions are repeated for each desired surface time-of-flight, starting at the maximum distance between specimen and transducer. Thereby, it is possible to adapt the permanent signal after each change of the measurement setup to compensate time shifts, caused by temperature changes.

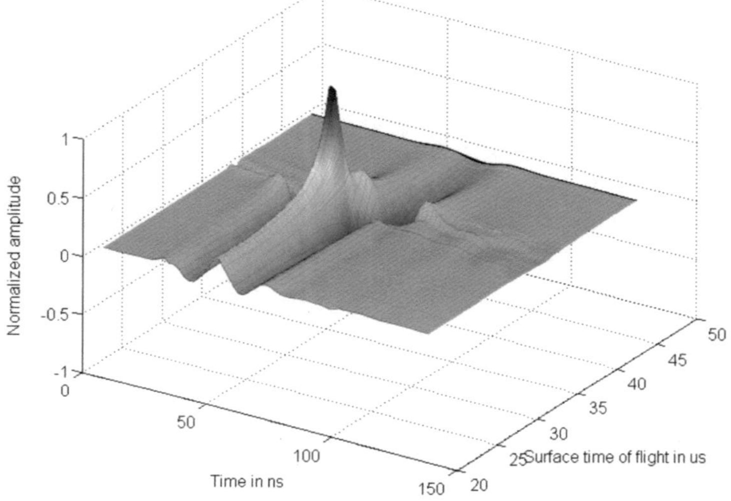

Figure 19: Reference signal in dependence of surface time-of-flight *TOF* (focus position); 75 MHz transducer,

The processed reference signals for the 75 MHz transducer are shown in figure 19. They contain the signal shape information for a wide focusing range. For a homogeneous media, like water, the corresponding *GSP* value can be determined with: $GSP = TOF \cdot c_W^2$. The signal amplitude depends on the focus position and the attenuation. Both dependencies can be compensated if necessary.

3 Ultrasonic Simulation

3.1 State of the art

There are many different approaches for an ultrasonic simulation, each with different assumptions, computation time and precision. The main categories are analytic methods, finite element methods (FEM), finite boundary element methods (BEM) and ray trace methods. All can be used in hybrid models.

Finite element simulation methods are based on analytical assumptions for very small spaces within a discretized media. With a transient simulation, the results are determined directly from the excitation signal(s). A separate determination of the impulse response and the convolution with the excitation signals(s) is not necessary. The results are generally very realistic as long as the discretization is fine enough for the used frequencies (numerical dispersion) [5]. It is limited to small dimensions and short simulation time due to the growing computational effort. It is often used for small local problems like the prediction of the sound field and wave types for certain formations, like the surface wave investigations in [6]. For larger formations, the mathematical effort can be reduced with symmetry assumptions and hybrid models with other simulation approaches like a combined FEM-BEM in [7] and [8]. A good overview of numerical simulation techniques is presented in [9].

For the Elastodynamic Finite Integration Technique (EFIT) the material parameters and the equations are discretized on a staggered grid [10]. This reduces the computational effort and allows more complex simulations. More optimizations are possible for symmetric problems. An axial symmetric 3D simulation can be transferred to a 2D simulation (CEFIT) as demonstrated in [4] and [11]. For scattering inhomogeneous materials an average 2D simulation can also be used [12][13]. With a 3D simulation the results are more realistic but much more expensive.

Simulations with ray tracing algorithms can be very fast compared to other algorithms if the specimen is not very complex, as in [14]. The effort grows exponentially with complexity, and the proximity to reality mainly depends on the assumed boundary conditions. Often planar waves and interfaces are assumed like in [15] and [16] to simplify those conditions. For a tilted wave front the boundary conditions become more complex and analytically solutions for the reflection and transmission coefficients can still be found. Simulation of frequency dependent attenuation is also possible [17]. Average material properties are assumed in reference [16] for sound propagation simulations in highly complex composite materials. This approximation is mainly true for the primary wave front but not for reverberant waves.

Analytical sound field simulations, as with the Helmholtz equation in [18], are as exact as the underlying assumptions. It is possible to get exact equations for simple geometrics and boundary conditions, but for realistic specimens a discretization is often necessary. Analytical methods are very useful for hybrid or semi-analytical approaches [19]. Large homogeneous blocks as in [20] are calculated much faster than with numerical methods. The integral transformation can reduce the number of dimensions in layered structures, what decreases the mathematical effort [2].

3.2 Simulation approach

The simulation of the ultrasonic analysis system should be very accurate and comparable to a measurement. This is necessary for the reconstruction algorithms with direct comparison to simulated results and high quality ultrasonic simulations. For this, the simulation requires a sound field measurement of the transducer used in the ultrasonic system as input. In contrast to parametric sound field approaches, the sound field shape and the signal are not idealized. This avoids systematic discrepancies between measurement and simulation.

One major disadvantage of sound field measurement as input for simulations compared to parametric approaches is the noise of a sound field measurement, which reduces the quality of simulations and causes artifacts. Noise can be minimized by increasing measuring time and by using newly developed filters. Generally the noise of reflection signals from wider (reference) reflectors is much lower than the noise of point reflector signals from sound field measurements. For this reason, two basic simulation approaches have been developed: the sound field measurement based simulation and the reference signal based simulation.

3.2.1 Sound field measurement based simulation

The sound field measurement based simulation is directly initialized with the sound field measurement results. The transient pressure signal in water from the incoming wave at each r position (cylindrically symmetric system) and a fixed z level is used as given input for the sound pressure in the pulse excitation phase. In the reception phase the signals can be recorded at any point of the system. For an impulse-echo system the received signal is obtained from the convolution of the input pressure signals s_{ex} and the received pressure signals s_{re} at the same points. The integration over the sound field measurement radius R

delivers the electrical signal s_r:

$$s_r(t) = k_T \cdot \int_{r=0}^{R} 2\pi r \cdot \int_{\tau=0}^{T} s_{ex}(r, t+\tau) \cdot s_{re}(r,t) d\tau dr \,, \tag{28}$$

where k_T is the conversion factor of the transducer and T the observation time period. In an one transducer system, s_{re} is determined by the convolution of s_{ex} with the impulse response of the specimen *SP*.

The sound-field measurement approach is well suited for finite element simulations (e.g. EFIT) and semi analytical simulations. The propagation of the sound waves in the medium can be simulated to investigate and to understand propagation effects.

The major challenge is the high sensitivity to noise from the sound field measurement. Due to the integration of large noisy outer areas in equation (28), the noise is amplified. Proper filtering with an optimal relation of information preservation and noise removal as in figure 6 and 8 is necessary. Too restrictive filters would cause less realistic results and too soft filters cause stronger artifacts due to noise. This approach allows a very realistic simulation in general but with high efforts for a high quality sound field measurement.

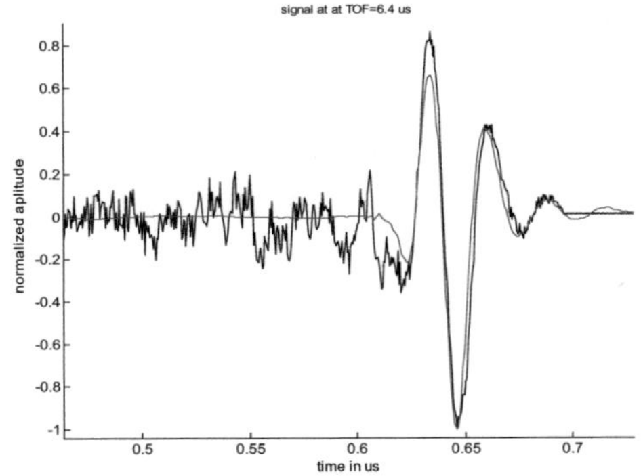

Figure 20: Comparison of a measured signal (grey) and a calculated signal from a sound field measurement (black)

Some sound propagation simulations with different simulation algorithms and a sound field measurement as input are shown in figures 14 and 15. A received electrical signal, which is obtained with equation (28), is displayed in figure 20. It is compared to a directly measured signal, which shows much less noise in the beginning phase of the pulse. Most of noise comes from the wider outer areas of the sound field measurement, where the minimal phase criterion for the self-deconvolution can not be fulfilled and artifacts arise. Due to the higher signal amplitude in the later phase of the pulse, the minimal phase criterion is fulfilled better and the artifacts caused by noise are much lower.

This type of simulation is well suited for the determination of the sound field and its propagation. In contrast to that, the quality of received signal is relatively low. For some applications it might be sufficient to reduce the noise with restrictive filters.

3.2.2 Reference signal based simulation

The reference signal based approach uses the relatively strong reflection signals of a plane surface as input for the simulation. Those reference signals already contain all interference effects of the sound waves at a plane material interface in dependence of the sound field position. Integration is unnecessary, thus there is little sensitivity to measurement noise. There is a systematic discrepancy caused by the changing interference effects for different material setups. This also leads to the limitation of orthogonal plane material interfaces like the reference, because other shapes would cause different interferences. The main advantage of this approximation is that it is well suited for ray-trace algorithms for layered specimens. It provides the ability for very fast simulations of plane interfaces. The sound field is decomposed into plane waves and the signals are assumed to be the same as the reference signals. The amplitude A_{ni} of each received echo is determined from the spherical interference approximation with equation (25). Since the pulse signals at the interfaces are known, the simulation algorithm only has to determine the impulse response $SP_i(t)$ of the specimen, and this is convoluted with the normalized reference signals $rs_n(GSP_R,t)$. The reflection signal s_r is obtained by the summation over all used initial angle θ_0 indexes i:

$$s_r(t) = \sum_i \int_{\tau=0}^{T} SP_i(t+\tau) \cdot rs_n(GSP(t-\tau),t) \cdot A_{ni}(GSP(t-\tau))d\tau \,. \tag{29}$$

This approach is very effective and enables very realistic and fast simulations for layered structures. In general, it is less realistic than a sound field measurement based simulation, but due to the much lower reference noise, it can be still more realistic.

This reference signal based ultrasonic simulation approach is used for reconstructive signal analysis of this thesis. It can be used for direct comparison of measured reflection signals with simulated ones as shown in figures 26 and 45. The signals show a high conformance compared to the sound field measurement based approach.

3.3 Determination of the impulse response

The impulse response of an ultrasonic simulation describes all propagation effects of the complete ultrasonic system. It is independent of the actual signal shape that is emitted from the transducer. If the impulse response SP is known, the received signal can be obtained from the convolution of the impulse response with the excitation signal and the transducers impulse responses as in equation (1). The impulse response has to be determined from the integration of the transfer functions between each point of the excitation transducer and each point of the reception transducer (Green's functions). Since this requires a full analytic solution of a three-dimensional problem, simplified assumptions are used to solve complex systems efficiently under practical conditions. One of those general assumptions is rotatory symmetry, which reduces the effort to a two-dimensional problem and is valid for most transducers of ultrasonic microscopes. The symmetry also requires a rotatory symmetric sound path and specimen. So it is not possible to simulate a three-dimensional specimen with edge effects at non-planar material interfaces. Additional assumptions, like one-dimensional wave propagation with two-dimensional effect considerations, depend on the algorithm used.

3.3.1 1D ray-trace algorithm

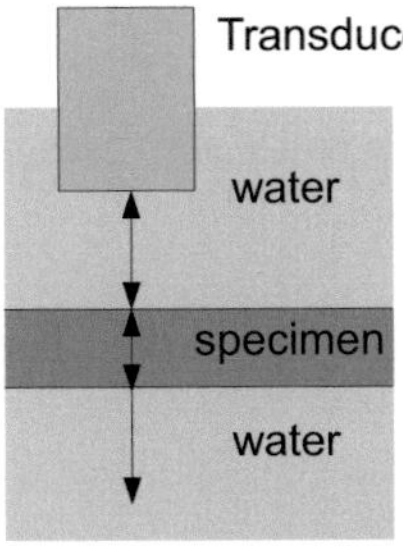

Figure 21: 1D ray trace model

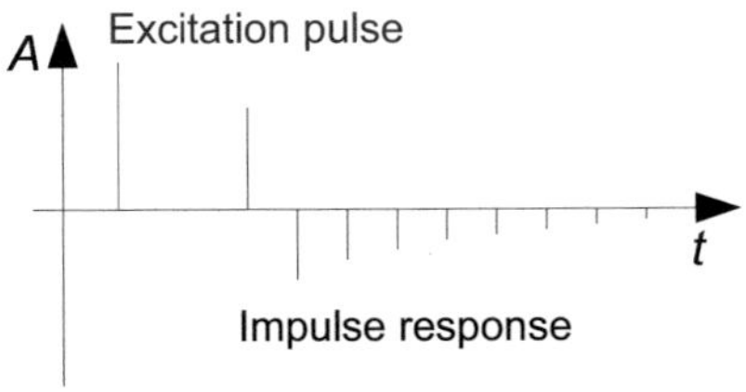

Figure 22: Impulse response according to figure 21

For the one-dimensional ray trace algorithm all acoustic waves are assumed to be planar and with orthogonal incidence to all interfaces. Since the sound coupling is done with a fluid, like water, there are no shear wave modes. A ray represents the wave normal in the propagation direction (figure 21). According to the reflection and transmission coefficients from the equations (7), each ray splits up into a transmitted and a reflected ray at a material interface. If we assume the excitation of a Dirac impulse ray from the transducer, the received reflection signal at the transducer will represent the impulse response (figure 22) of the system. Each peak of the impulse response represents a path of the excitation ray and has a total time-of-flight *TOF*:

$$TOF = \int_S \frac{ds}{c(s)} = \sum_{n=1}^{N} \frac{d_n}{c_n}, \tag{30}$$

where S is the sound path, c the sound velocity, N the number of layers in the sound path and d the thickness of each layer. The amplitude A of each peak (ray) is determined by the initial amplitude A_0, the reflection R and transmission T coefficients in the sound path and the idealized (no frequency dependency) acoustic attenuation a [dB/m]:

$$A = A_0 \left(\prod R\right)\left(\prod T\right)\left(\prod_{n=1}^{N} 10^{\frac{a_n \cdot d_n}{20\text{dB}}}\right). \tag{31}$$

In addition to the one-dimensional impulse response, the generalized sound field position *GSP* from equation (15) is required for the determination of the corresponding reference signal for each ray, which contains the two-dimensional sound field information like focusing:

$$GSP = \int_S c(s)ds = \sum_{n=1}^{N} c_n d_n . \tag{32}$$

This reference signal is used for convolution to obtain the received signal. Thus amplitude considerations in dependence of the focus position are simultaneously included. Since all rays are infinite, they have to be canceled when they leave the fixed time window to ensure a finite processing time.

The one-dimensional ray trace algorithm is convenient for the simulation of specimens, which consist of similar soft materials, where almost no shear waves are generated by mode conversion. The discrepancy between simulation and real measurement increases with the use of harder materials, which cause stronger mode conversion at the material interfaces. A comparison between simulation and measurement for a single glass layer is shown in figure 23. Glass is relatively hard and strong mode converted echoes occur, which are not visible in the simulation. Only the pure pressure wave echoes are simulated. The "phantom" echos directly behind the surface- and other strong echoes are no mode converted echoes and belong to the strong echo in front. Since reference signals are used for the convolution, all effects from the transducer design, are well reproduced.

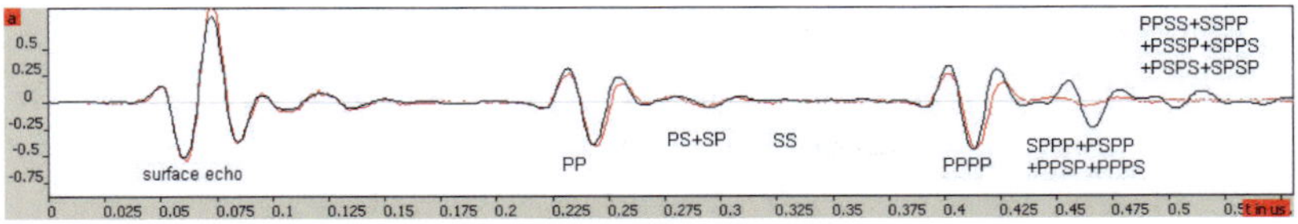

Figure 23: A-Scan comparison of a 515 um SiO_2 layer in water

75 MHz transducer ; black: measurement, red: simulation with 1D ray-trace

3.3.2 2D ray-trace algorithm

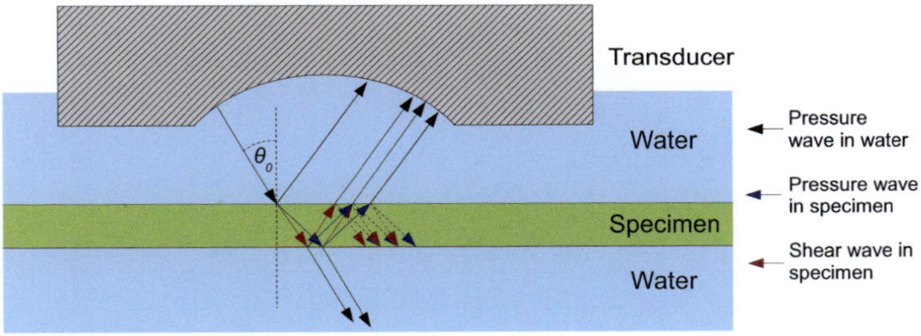

Figure 24: 2D ray trace model

The two-dimensional ray trace algorithm assumes plane waves and plane material interfaces. In contrast to the 1D algorithm the wave normal (ray) does not have to be orthogonal to the interfaces (figure 24). At the interfaces the acoustic waves are refracted according to Snell's law in dependence of the angle of incidence θ_0. Through mode conversion, shear waves are generated from pressure waves and vice versa. At a solid-solid material interface, each incoming wave splits up into four waves: a reflected shear wave R_S, a reflected pressure wave R_P, a transmitted shear wave T_S, and a transmitted pressure wave T_P. Their amplitudes can be calculated with the reflection and transmission coefficients from the equations (6). Since the sound path is not orthogonal to the layers, the effective thickness of one plane layer in equation (30) to (32) is:

$$d_n = \frac{d_{\perp n}}{\cos(\theta_n)}, \text{ with } d_{\perp n} \text{ as orthogonal thickness.} \qquad (33)$$

A real sound field of a spherical focusing transducer is very much alike a small section of a spherical wave front. The component of the sound field with the highest energy has an angle of incidence that is above zero (figure 10). This causes mode conversion at the material interfaces and this effect should be considered with the two-dimensional ray-trace simulation algorithm.

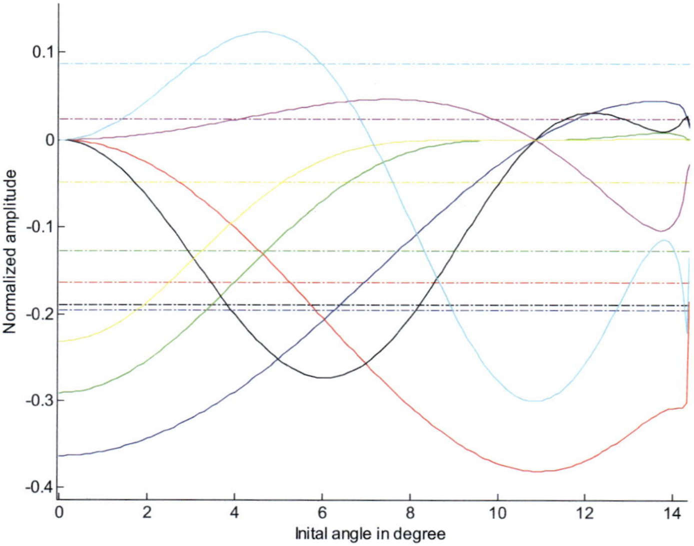

Figure 25: Reflection amplitudes from a single 1 mm glass slide (water-glass-air interface); plane waves; slash-dot-lines: focused echoes from a measurement with a 75 MHz transducer (this is the "goal" and is no angle dependent value); solid lines: 2D single ray-trace-simulation focused echoes in dependence of the initial angle; blue: PP-reflection; red: SP+PS reflection; magenta: SS-reflection; green: PPPP-reflection; black: SPPP+PSPP+PPSP+PPPS-reflection; cyan: SSPP+PPSS+SPSP+PSPS+PSSP+SPPS-reflection; yellow: PPPPPP-reflection

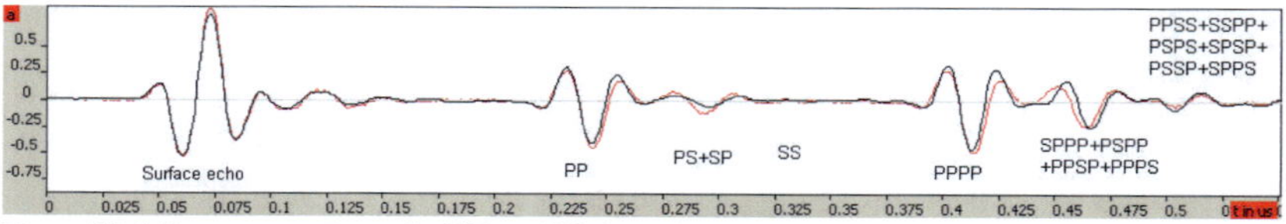

Figure 26: A-Scan comparison of a 515 µm SiO_2 layer in water 75 MHz transducer ; black: measurement, red: simulation with sound field decomposition

Figure 25 displays the maximal amplitudes (usually in focus) of the first reflections from a 1 mm glass slide. The measured maximal amplitudes are compared with results of a 2D ray-trace simulation of a single initial ray in dependence of its initial angle θ_0. There is no initial angle, where all calculated reflection amplitudes meet the measured ones well. There is some conformance in the region between $\theta_0 = 4°$ and $\theta_0 = 6°$, but the relative error is still above 10%. One major reason for this is that the average effective angle does not correspond to

Snell's law after mode conversion, because wave components with higher angle have higher conversion between pressure and shear waves.

All approaches with one initial ray for the simulation of ultrasonic wave propagation were designed to provide relatively good results for low effort. But the precision is not sufficient for the high demands of reconstruction algorithms and simulation investigations. Therefore a new approach is developed, which is based on the decomposition of the sound field into governed plane waves with equation (8). Each plane wave (ray) has a different initial angle and an initial signal amplitude according to the sound field measurement (figure 10). For each initial ray a separate simulation is set up and the single results are superimposed to obtain the received signal with the sound field modeling. The amplitude of the received rays is determined from the initial amplitude, the reflection and transmission coefficients and the attenuation along the ray path as in equation (31). Additionally, the amplitude of the reference signals must be normalized, because interference effects are now modeled with the normalized amplitude factor A_{ni} from equation (25) in dependence of the *GSP* value of the received ray. This approach can also be used in conjunction with the sound field measurement signals instead of the reference signals (e.g. figure 14). The main advantage would be more realistic interference effects. Artifacts from the higher signal noise of the sound field measurement can reduce this advantage.

The number of rays used for the simulation is not fixed and can be varied depending on the precision demands. In this thesis, an initial angle spacing of 1° has been chosen. It is a good compromise between precision and simulation speed. Based on the sound field measurement (figure 10), a total number of 25 rays has been used for the simulations. A comparison between measurement and simulated reflection signal is shown in figure 26. The two signals are highly correlated and all strong reflection modes have similar amplitudes. There are some small pulse shape changes, which are partially caused by frequency dependent attenuation. Such effects are not considered in the simulation jet.

3.3.3 Complexity reduction – optimizations

The efficiency of ray trace algorithms depends very much on the number of layers and the simulation time. An approximation of the total number of different ray paths N_R in the total simulation time t_{total} is:

$$N_R \approx b^{\frac{t_{total}}{t_S}(N_I - 1)}, \tag{34}$$

where b is the average number of resulting rays at an interface with a value between 2 and 4, N_I the number of interfaces of the specimen and t_S the average time-of-flight through the specimen for shear and pressure waves. For specimens with several layers, N_R may easily grow beyond the number of calculations necessary for discrete numeric simulations.

One possible approach for the reduction of complexity would be the cancellation of rays with very low intensity. After interaction with several interfaces, the amplitude of single rays is usually very low and it does not cause significant signal content. The precision of the result would not be negatively affected, as long as the cancellation threshold is below the noise level and the pulses do not superimpose.

Once a high number of rays are received at the same time, extremely small amplitudes may still cause significant signal content. This can be the case at a specimen with a lot of layers. Even if one single ray has a relative amplitude of 10^{-6}, one million of those rays from multiple reflections may sum up to an amplitude close to 1. Under those conditions, the ray cancellation at 0.01% of the initial amplitude for an acceptable processing time causes wrong results. Effects from deep layers are missing (figure 27b), although the measurement provides strong signals from all layers of the eight-die-stack (figure 27a). Choosing a lower cancelation criterion rapidly increases the computation time. An additional error effect comes from the limited numerical precision when adding very small and large values at the receiver. Very weak waves can get numerical “lost”.

A solution for this problem can be achieved with a new approach that determines a low-pass filtered time discrete impulse response function. The here introduced Local InterAction Ray Trace (LIART) algorithm combines rays at same times and places. It is based on the same equations as the ray trace algorithm, but it calculates time discrete transfer functions between the interfaces instead of the calculation of single rays. Those transfer functions are applied to the incoming signal content A of all rays within each time step $_\Delta t$ and at each interface i, for the forward and backward direction separately:

$$A(i,t) = T_i \cdot R_i \cdot 10^{\frac{a_i \cdot d_i}{20\text{dB}}} \cdot \int_{t-t_i-\Delta t/2}^{t-t_i+\Delta t/2} A(i-1,\tau)d\tau \text{, with } t = n\,{}_\Delta t,\ n = \text{sample and } t_i = d_i / c_i. \quad (35)$$

It can be initialized with the initial ray (pure LIART) or with the weak canceled rays of the ray trace algorithm (hybrid LIART). Since each ray carries the information of time, amplitude and *GSP*, it is not possible to sum up rays exactly without loosing information and with effective complexity reduction. Rays at the same time and place may have a different

GSP value. Normally, those *GSP* values are not very far off and they can be averaged with amplitude weight:

$$GSP(i,t) = t_i \cdot c_i^{\,2} + \frac{\int_{t-t_i-\Delta t/2}^{t-t_i+\Delta t/2} GSP(i-1,\tau)\cdot\left|A(i-1,\tau)\right| d\tau}{\int_{t-t_i-\Delta t/2}^{t-t_i+\Delta t/2} \left|A(i-1,\tau)\right| d\tau}. \qquad (36)$$

This averaging effect can be minimized by using the hybrid LIART, where only the very weak rays are calculated with the LIART and the strong part is calculated with ray trace. Both results are convoluted separately and summed up. The total number N_L of calculations for LIART is:

$$N_L = \frac{2 \cdot b \cdot N_I \cdot t_{total}}{\Delta t}. \qquad (37)$$

The effort is now linear with the number of interfaces N_I in the specimen and the simulation time. This algorithm requires a time step Δt shorter then the minimal time-of-flight between two layers and the sampling theorem must be fulfilled for the used reference signals. For reference pulses, Δt should be smaller then one tenth of the average time period $(1 / f_c)$.

Results of the LIART algorithm in comparison with the normal ray trace algorithm and measured data are shown in figure 27. The LIART results show the same effects from the flaws of each layer, like the measured data. Simulated and measured Signals of the same defect type have a correlation of 95%, where 2.5% mismatch is already caused by signal noise. The computation time for the simulation of one reflection signal is now below one second, whereas it has been several minutes for the ray trace algorithm with ray cancellation.

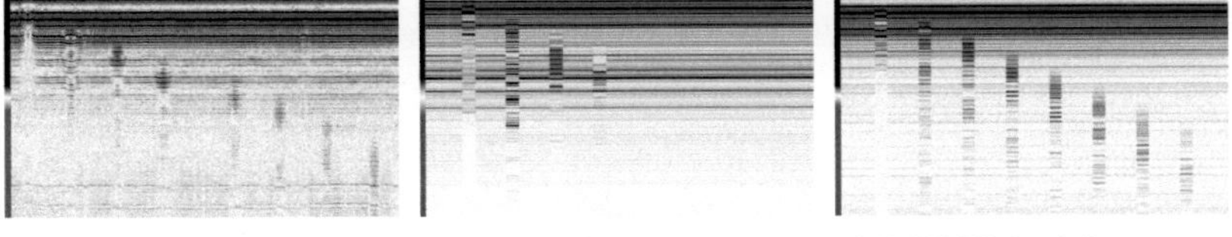

a) measurement **b) 2D ray trace simulation with ray cancelation at 0.01% of the initial amplitude** **c) 2D LIART simulation**

Figure 27: Comparison of measured and simulated B-Scans from reference specimen 3 eight-die-stack with air inclusions at each layer (see appendix pages 97-99); 75 MHz transducer; scan area: 17.5 mm x 1.1 µs

4 Deconvolution – Determination of reflection parameters

Ultrasonic reflection parameters can be determined directly from the ultrasonic signal, as long as the single echoes are not overlapping. If the echoes are overlapping, they have to be separated by deconvolution, to obtain the reflection parameters. Deconvolution is defined as the inverse operation of the linear convolution. For ultrasonic systems, a linear system is assumed in general. Under the assumption of plane waves with orthogonal incidence in a layered structure, the wave propagation is simplified to an one-dimensional problem. The reflected or the transmitted signal $s_r(t)$ can now be determined from the one-dimensional version of equation (1), which is the convolution of the reference signal *rs* and the impulse response *h*:

$$s_r(t) = \int_{\tau=-\infty}^{\infty} rs(\tau) \cdot h(t-\tau) d\tau \,. \tag{38}$$

The reference signal *rs* is determined from the convolution of the excitation pulse with emitting transducer impulse response, the receiving transducer impulse response and the receiver circuit impulse response. *rs* is equivalent to a reflection from an orthogonal plane interface. The impulse response *h* is determined by the layer structure of the specimen and represents the material characteristics. It is directly related to the reflection parameters, which are represented by the time and the amplitude of each reflected or transmitted pulse. As long as the time between two pulses is greater than the pulse length, those parameters can be determined directly form the received signal. Otherwise, the pulses are superimposing and have to be separated by deconvolution.

Generally, we can measure the received signal s_r and the reference impulse *rs* directly. Since both signals have systematic and random noise components, it is not possible to perform an exact mathematical deconvolution. The deconvolution algorithm has to be insensitive to noise, although noise can be partially removed with well-designed filters.

Under realistic conditions with spherical focusing transducers, rs cannot be represented by one unique signal. It depends on a variety of factors such as sound field position and frequency dependent attenuation, which are different for each received pulse. Therefore, the deconvolution algorithm must predict the signal $rs(\tau, t)$ in dependence of the time-of-flight t and has to be capable to handle a varying reference signal *rs*.

4.1 State of the art

Many algorithms are available for the advanced evaluation of the reflection signal. They estimate the parameters of the single echoes and can be divided into two main categories: decomposition and deconvolution techniques.

4.1.1 Decomposition techniques

Decomposition techniques search for a representation of the original signal with a sum of analytical wavelets. Those wavelets can be of any shape. The most famous example is the Fourier transform. For the ultrasonic analysis, those wavelets should represent single echoes to obtain the reflection parameters. Because of superimposed noise on a measured ultrasonic reflection signal, an exact representation is not always possible or useful. Thus, several presumptions are used to achieve good and useful approximated representations.

The matching pursuit (MP) algorithm [21] decomposes an arbitrary signal into a linear superposition of waveforms. They are selected from a redundant dictionary of functions and are chosen in order to best match the signal structures. In [21] it is mainly used for speech recognition with a dictionary of Gabor functions as an adaptive time-frequency transform. It can be used for the decomposition of audio signals with good success [22] and a simple note detection algorithm that also works with overlapping notes is demonstrated. In [23] the matching pursuit algorithm is used for angle of arrival detection and its superiority is shown with a comparison to a simple beam-former.

After decomposition, the signal can be represented with a few terms of Gabor functions. The error to the original signal is mainly noise. Thus the MP is very good for noise suppression and data compression [24]. There are special extensions [25] to the MP to make it more flexible and optimized for a special type of signals. Reference [26] gives an approach for an error estimation of the MP and optimization. It can be used for ultrasonic signal decomposition, resulting in a high-fidelity time-frequency representation [27].

In most cases Gaussian pulse decomposition algorithms suit the nature of ultrasonic reflections very well. The parameters of each pulse are amplitude, arrival time, center frequency, bandwidth and phase. Those parameters can give information about frequency dependent absorption, scattering and dispersion effect, and the geometric shape, size and orientation of reflectors. In reference [28] a maximum likelihood model of the backscattered echoes was developed to estimate all unknown parameters. The convergence problems due to local minima have been minimized by a two-stage iterative Expectation Maximization

(EM) algorithm. The first stage estimates the expected signal and the second stage estimates parameters with the maximum likelihood criterion. The result is less dependent on the initial guess and the number of echoes is fixed. Reference [29] models the ultrasonic reflection signal as a superposition of many Gaussian echoes corrupted with measurement noise. The parameters from the pulses are estimated with the Space Alternating Generalized Expectation Maximization (SAGE) algorithm. It is similar to the EM algorithm but with faster convergence and a Gauss-Newton algorithm to handle white noise. The demonstrated examples are relatively simple. If the ultrasonic pulses are similar to a Gaussian pulse, the SAGE algorithm can be used for deconvolution problems with interfering echoes [30]. It resolves the single echoes with a spacing of approximately one wave length from a real specimen. Reference [31] contains a comparison to other decomposition algorithms and a very good summary of the SAGE algorithm. Some good practical results of nondestructive evaluation applications for reference [31] are shown in [32]. The Gaussian pulse decomposition algorithm can also be used for a high quality time-frequency representation of the ultrasonic signal which is very robust to noise [33].

The wavelet transform was first proposed for the analysis of seismic data in reference [34]. It is an integral transform with one wavelet of constant shape. This “mother” wavelet is shifted and scaled to “daughter” wavelets which are used for the transform and that are represented by the wavelet coefficients. Each coefficient stands for the magnitude of a wavelet for a time and a frequency scaling factor. In reference [35] it is compared to the short-time Fourier transform for a time-frequency analysis and localization. Paper [35] contains a very comprehensive theory overview for the wavelet transform. Applications for the wavelet transform are noise resistant pulse detection [36], frequency-time-investigations [35][36], data compression [37][38], noise reduction [37] and signal deconvolution [39]. The ultrasonic decomposition techniques are limited to the signal properties of the wavelet used. They work well if one single echo is similar to a Gaussian pulse [37][38]. Wavelet transform is widely used for image processing and deconvolution as in reference [39]. Theoretical issues for an optimal wavelet deconvolution are studied in reference [40].

The chirplet transform used in reference [41] decomposes any ultrasonic signal into the superposition of multiple single chirplet echoes. It can be used to obtain a time-frequency representation, the echo amplitude, time of arrival, center frequency, bandwidth, phase, and chirp rate of multiple interfering ultrasonic echoes [41].

4.1.2 Deconvolution

The Deconvolution is a special case of the decomposition where a wavelet represents a single echo. Usually this reference echo and the reflection signal are measured signals and superimposed by noise. Thus this inverse problem has no explicit solution. There are many algorithms available to find an optimal solution depending on presumptions.

An overview of earlier deconvolution techniques like Wiener-filters and spiking deconvolution is presented in [42] and [43]. The results are not satisfying and they are often used as "bad" examples. Derivatives like spectral extrapolation and L2 have better results [43]. They are optimized for low computational complexity. The Van-Cittert deconvolution [44] is a very simple process that should lead to a good solution after a certain number of iterations. First investigations showed that it is very unstable and can lead to an oscillating solution with infinite magnitude. A comparative study of blind deconvolution, where no reference impulse needed, is presented in [45]. They are based on statistically data of the pulses. The results are not very useful for this thesis, because they contain many artifacts.

Well trained neural networks can solve almost any problem. When all representative situations are trained, it can be used for the deconvolution of ultrasonic data [46]. In other cases an unbiased maximum likelihood method would produce more reliable results.

Due to noise and bandwidth limited signals, the direct deconvolution is very unstable and produces unusable results [47]. It is usually performed by the division of the FFT from the measured signals like in [48]. A time-domain matrix solution [49] would lead to similar results. With the right type of filter the result is much more usable but bandwidth limited. It is also possible to apply a filter to the excitation pulse to eliminate the impulse response of the transducer [47]. The main challenge is to find an optimal filter depending on the application. One iterative approach for the filter parameter determination is demonstrated in [50] and in [51] with improved results. More promising is a hybrid Fourier-wavelet-filter as proposed in [52].

A maximum likelihood estimation algorithm is presented in [53]. It is a tree-based algorithm with reduced complexity proportional to 2^(number of samples of the reference impulse). The results are very good but it requires a very short reference impulse to be computationally solvable. To reduce the computational effort, a Bernoulli-Gaussian process can be assumed for the pulse positions [54]. It is a statistical approach that does not consider all possibilities but produces good results. In reference [55] it is extended to handle phase shifts and colored noise. Under given limits, the reference wavelet and the noise properties can also be estimated by statistics [56]. The obtained numerical results are very promising. Some more optimizations have been done in [57].

Deconvolution is also possible with model based approaches, which compare the measured signal to the simulated result of a model. Those models are generally restricted to simple structures and one-dimensional assumptions as in [58]. This approach can also be used for reconstruction algorithms.

A potentially very flexible deconvolution and decomposition algorithm is described in [59]. It is based on a correlation analysis with an arbitrary reference signal, in which the measured signal is decomposed. The result represents the approximated impulse response. The sensitivity to noise and disturbances is relatively low and it can easily be extended to handle varying reference signals. This algorithm is used and extended in this thesis.

4.2 Analytic signal investigations for deconvolution

Figure 28 shows an example Z-Scan impulse diagram of a glass slide in water. It contains all occurring reflection modes (e.g. surface echo, longitudinal backwall echo PP, single mode-converted shear wave backwall echo PS and SP, ...) in the displayed time range in dependence of the surface echo time-of-flight. The displayed data is the color coded envelope function (absolute Hilbert transform) of the reflection signals. The x-axis of the figure represents the time-of-flight relative to the surface echo and the y-axis of the figure represents the total time-of-flight (*TOF*) of the surface echo, which is sound velocity proportional to the z-position of the transducer and the focus position.

The goal of the deconvolution is to separate the reflections and to obtain single spikes for each echo like an impulse response (figure 22). Artifacts of the deconvolution result are mainly influenced by noise and the correctness of the reference impulse shape. Therefore, the shape of the single reflection modes is now compared with other reflections in dependence of the focus position. The normalized cross-correlation is an expression for the uniformity of two signals, where a value of one means perfect equality, zero no similarity and minus one perfect negative equality.

Figure 29 displays the normalized absolute correlation matrix of the surface echo with itself for different focus positions. As expected, the correlation of echoes with the same *TOF* (z-location of the transducer) is perfect. The signals of the 75 MHz transducer are highly correlated if both echoes come from an interface in front of the focus position (around $TOF = 35\ \mu s$) or from behind. A deconvolution of a signal from behind the focus position and a reference signal from in front of the focus position would cause 1 - correlation ≥ 10% artifacts. Unfortunately, the correlation of echoes directly in focus is very low, due to the phase inversion. Therefore, a high quality deconvolution requires out of focus measurements or a very high quality reference impulse prediction.

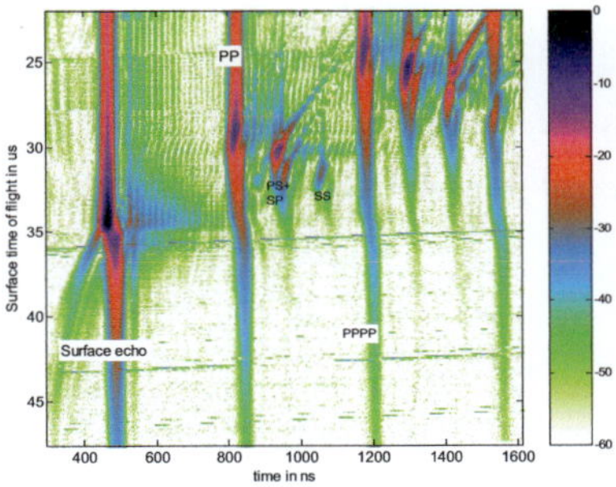

Figure 28: Z-Scan of a 1 mm glass slide 75 MHz transducer; reflection signal in dependence of the surface echo *TOF*

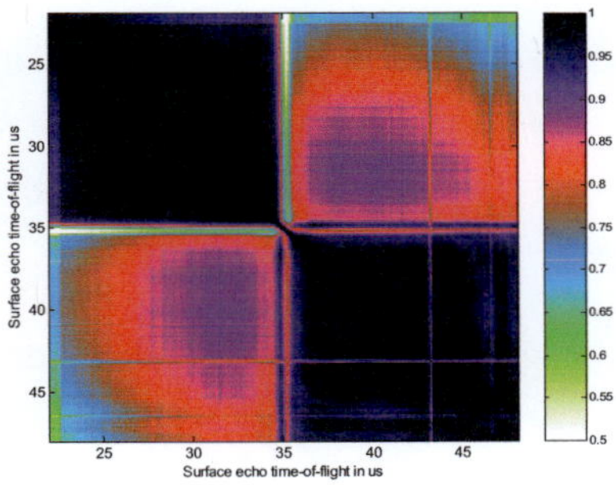

Figure 29: Absolute correlation of the surface echo with itself in dependence of the *TOF* (*z*-position of the transducer)

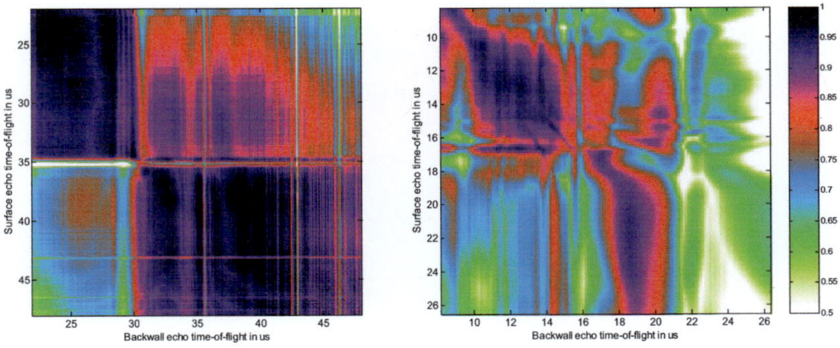

a) 75 MHz transducer, 1 mm glass slide

b) 15 MHz transducer, 1 mm plastic slide

Figure 30: Absolute correlation between surface echo and longitudinal backwall echo PP in dependence of the *TOF* (*z*-position of the transducer)

A similar behavior can be observed for different echoes (figure 30a). The focus time-of-flight of the backwall echo is shifted according to the *GSP* and the correlation characteristics are about the same. The data of figure 30b is from a 15 MHz transducer with a different design (no delay line), where almost only the pulses with the same *GSP* value are well correlated. This design difference causes stronger edge waves and thereby additional reflections from the edge waves. Now the shape of the overlapped reflections changes much more in dependence of the focus position, which reduces the quality of the deconvolution.

In general, a transducer with a shape stable signal (like the 75 MHz long delay line transducer) should be preferred for the deconvolution. As long as the attenuation is low, an arbitrary echo can be represented with a surface reflection with the same *GSP* value. Strong attenuation would cause a frequency shift and a different signal shape. Theoretically, it is possible to predict signal changes, if the frequency dependent attenuation is known. In any case, the focus position should be closely behind the layer of interest to obtain the best results.

4.3 Single reference pulse deconvolution

The basic linear deconvolution algorithm for ultrasonic reflections has been developed in reference [59]. It assumes a reflection signal composed of multiple superpositions of one reference pulse. One major advantage of this time domain based deconvolution algorithm compared to frequency domain approaches is the low sensitivity to noise. It does not require a restrictive frequency filter with information loss, so an optimal noise filter can be used without losing signal information.

The fundamental idea of the algorithm is to use the reference normalized cross-correlation $k(t)$ of the received signal $s_r(t)$ and the reference signal $rs(t)$:

$$k(t) = \frac{\int\limits_{\tau=-\infty}^{\infty} rs(\tau)\cdot s_r(t+\tau)d\tau}{\int\limits_{\tau=-\infty}^{\infty} rs^2(\tau)d\tau}, \tag{39}$$

to find the strongest occurrences of $rs(t)$ in $s_r(t)$ (figure 31). At the time t_{k_max} of the absolute maximum A_{k_max} of $k(t)$, $rs(t)$ is subtracted from the original signal $s_r(t)$ with the gain factor G:

$$s_r(t)_{i+1} = s_r(t)_i - G\cdot A_{k\_max}\cdot rs\left(t-t_{k\_max}\right). \tag{40}$$

The strongest echo is reduced to reveal weaker echoes and to separate superimposed echoes. In the beginning $h(t)$ is zero for all t. After each iteration loop i, the maximum is added to the impulse response $h(t)$:

$$h_{i+1}(t_{k\_max}) = h_i(t_{k\_max}) + G\cdot A_{k\_max}. \tag{41}$$

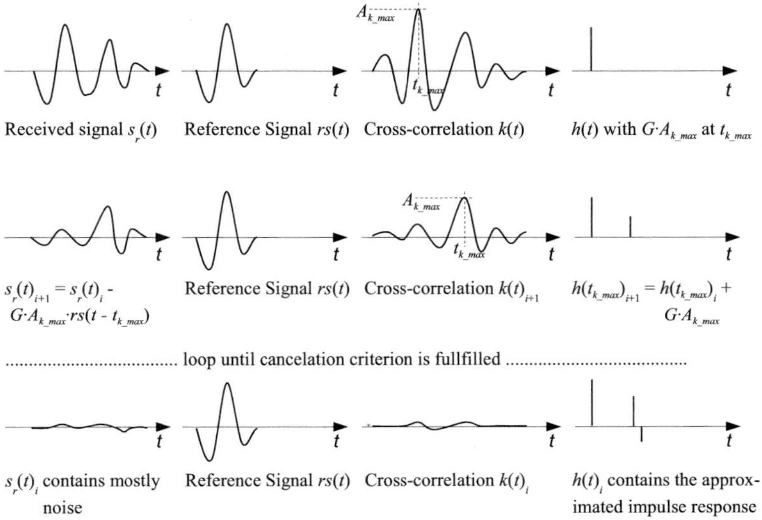

Figure 31: Iterative deconvolution scheme

For each loop, equations (39) to (41) are repeated until a minimum error criterion is reached or the maximum number of iterations is exceeded. The convergence of this algorithm is mainly influenced by the gain factor G, which should be between zero and one. A low value would cause a long processing time and high stability, and vice versa for a high value. For most applications, a value of $G = 0.5$ has been observed to be convenient. After the last loop, $s_r(t)$ contains almost only noise and $h(t)$ the approximated impulse response of the linear system (figure 31). The time resolution (distance between two separated pulses) depends on a variety of factors like the pulse spectrum and the amplitudes of the pulses. For an independent proper separation, the pulses have to be more than half the average wavelength apart, although the pulses can already be separated at more than half a wavelength of the highest frequency contents.

Artifacts of the deconvolution arise when the reference signal has a different shape than the pulses from the received signal. Their amplitude is directly related to the correlation factor, so a normalized correlation of 0.98 between received pulse and reference signal would cause about 2% artifacts. The correlation is also negatively influenced by the signal noise.

Figure 32 shows the deconvolution signal from the reference specimen 3 (see appendix pages 97-99). While it reveals the first three interface reflections very well, later reflections are too much superimposed and cannot be distinguished as clearly. In the raw reflection signal only the surface echo can be distinguished from the superimposed deeper echoes.

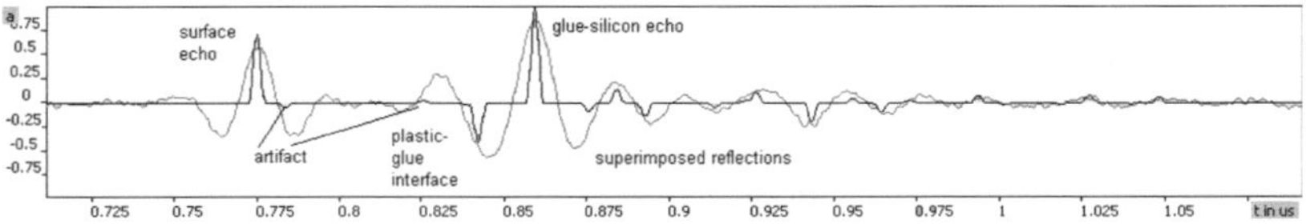

Figure 32: A-Scan from reference specimen 3 (no flaw position); 75 MHz transducer grey: measured reflection signal; black: low-pass filtered deconvolution signal

Based on the signal investigations for the deconvolution, we know that for certain transducers, like the 75 MHz one, all reflections from in front or behind the focus point are highly correlated (figure 30a). Under this condition, and with a focus point behind the signal of interest, the deconvolution with a matching single reference signal can produce good results. As a further development from [59], the reference signal is chosen according to the *GSP* value of the surface reflection.

For optimum results, the measured signal is processed with an optimal frequency filter that only reduces frequency components with higher noise content than the useful signal. The deconvolution algorithm is optimized to reduce the number of iterations necessary. Two pulses provide independent correlation results if they are more then one pulse length apart. This way, the number of required iterations is reduced by the number of pulse lengths that fit into the received signal length. After deconvolution, the result (impulse response) is represented by single spikes with the amplitude of the reference pulse at this position. Those spikes are not always very convenient for the ultrasonic analysis, because two very close spikes can represent one echo. This effect is avoided by the convolution of deconvoluted signal with a short Gaussian pulse (about 20% of the average wavelength), to obtain low pass filtered deconvolution signal (figure 32) for better processing.

The results for a single glass slide are shown in figure 33. As expected, the surface echo shows a clear positive peak with almost no artifacts, independent of the longitudinal focus position (vertical line). This is different for the negative longitudinal backwall echo PP, where strong artifacts occur (figure 33a and 33b) when the focus position is inside the specimen. In figure 33c the focus position is at the backwall. The artifacts are reduced, since all reflections in front of the focus position (figure 30) are highly correlated. The first mode converted negative shear wave echo PS+SP has almost no artifacts, since it is already in front of the focus point (figure 34).

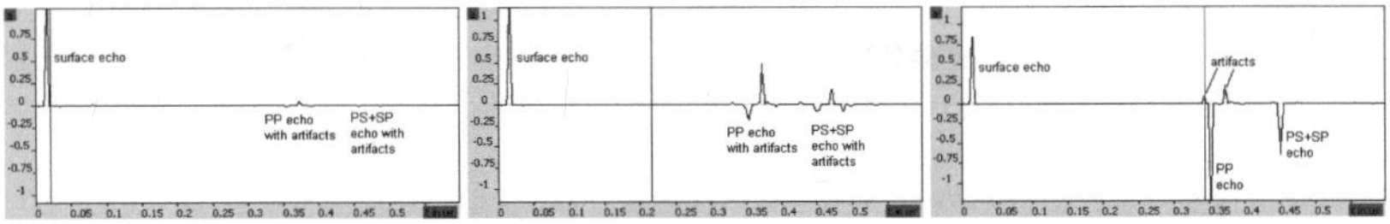

a) focus at surface **b) focus in center** **c) focus at backwall**

Figure 33: Deconvolution results from a 1 mm glass slide in water; 75 MHz transducer

4.4 Multi-pulse deconvolution

In a realistic environment the shape of each received pulse is different. The three-dimensional sound field propagation is transformed to an one-dimensional signal in the transducer. For each reflection or transmission mode the propagation and interference conditions are different, what causes changing pulse shapes. Under those conditions, the ultrasonic system can not be described exactly with an one-dimensional linear system.

A proper deconvolution algorithm has to be capable to handle a varying reference signal and to predict the signal shape of each received pulse. The here presented multi-pulse deconvolution combines all three-dimensional interference effects in the reference signals, and estimates the one-dimensional impulse response of a layered specimen. The time domain deconvolution algorithm from reference [59] is highly flexible and can be extended to handle a changing reference signal in dependence of the time-of-flight (pulse position in the signal).

All basic equations (39) to (41) remain valid, but now the reference signal *rs* also depends on the position of the echo inside the received signal. This position can be expressed with the *GSP* value for the corresponding reference signal. So $rs(\tau)$ is replaced by $rs(GSP(t), \tau)$ in equation (39) and (40). This enhancement of the normal deconvolution provides the ability to follow changes of the reference signal shape during the propagation process. The new challenge is now to predict the corresponding reference signal in dependence of the time-of-flight t. Due to possible superpositions of echoes, it is not possible to predict the reference signal directly from the received signal. The assignment of t and the *GSP* of a received ultrasonic signal is now introduced as *GSP* profile. Depending on the previous knowledge of the investigated specimen, it is possible to determine the $GSP(t)$ profile as shown in figure 34. It provides the corresponding time and *GSP* value for each echo ($*$ in the image) of a reflection signal. The different profiles are created for different analysis situations. Here "constant" stands for single pulse deconvolution, "homogeneous" is described in chapter 4.4.1 and "profile" in chapter 4.4.2.

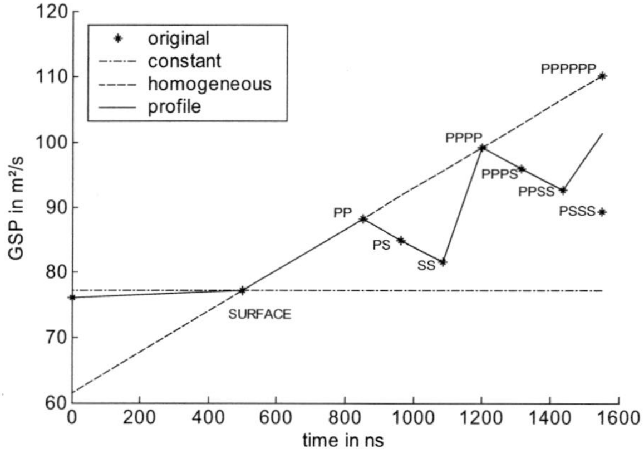

Figure 34: *GSP* profiles for a 1 mm glass slide in water

4.4.1 Homogeneous multi-pulse deconvolution

For the homogeneous multi-pulse deconvolution, a homogeneous material block with only one sound velocity is assumed. The $GSP(t_{SF})$ value for the surface echo is determined from the water delay line and all *GSP* values for other reflections are determined from this one on, with an average sound velocity c_{av} for the specimen:

$$GSP(t) = GSP(t_{SF}) + (t - t_{SF})c_{av}^{\;2}. \tag{42}$$

This prediction algorithm provides good results if the investigated specimen is of only one material or very similar materials. Figure 35 shows the results for a single glass slide. All longitudinal reflections are well deconvoluted with almost no artifacts, but the mode converted shear wave echoes contain a lot of artifacts due to the different sound velocity. This linear approach is very simple and a good compromise for many applications where the exact material structure is unknown and an average estimation is sufficient.

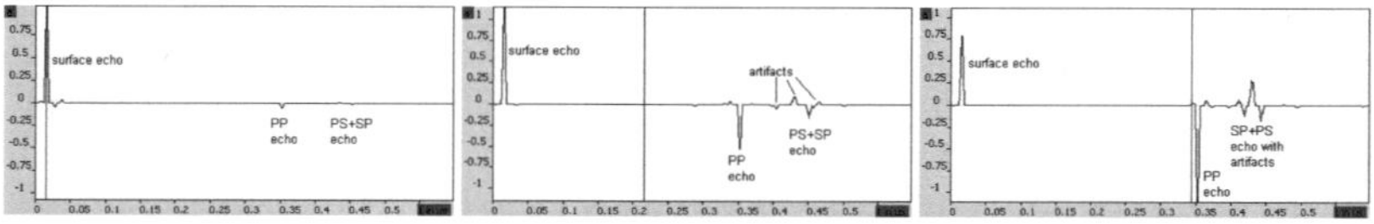

a) focus at surface **b) focus in center** **c) focus at backwall**

Figure 35: Homogeneous multi-pulse deconvolution results from a 1 mm glass slide in water

4.4.2 Multi-pulse deconvolution with simulated *GSP* profile

Multi-pulse deconvolution can use the *GSP* values from the impulse response of an ultrasonic simulation. The idea is an inverse approach, which assumes a known specimen structure and determines all occurring reflections with their amplitude and the *GSP* value. This information is used for the determination of the *GSP* value for all positions of the reflection signal. Since only a finite number of positions are defined by the received echoes, all values between two echoes are interpolated. A simple algorithm for this is the linear interpolation as shown in figure 34. It causes sharp edges at the points with a known *GSP* value, which could be reduced by a higher order interpolation. If two or more reflections arrive at the same time, the *GSP* value is averaged with amplitude weight at this position, like for the PPPPPP and the PSSS reflection in figure 34. In this case, the *GSP* values might not match and artifacts can occur. In all other cases, the *GSP* value is always correct as long as the simulated material structure matches the real structure precise enough. Small discrepancies of times can cause large changes of the *GSP* value. For this reason, this approach should not be used for unknown specimens. It is very convenient if the structure is determined within a reconstruction process, and provides high quality results with almost no artifacts for all strong echoes, including shear wave and multiple reflections, as in figure 36.

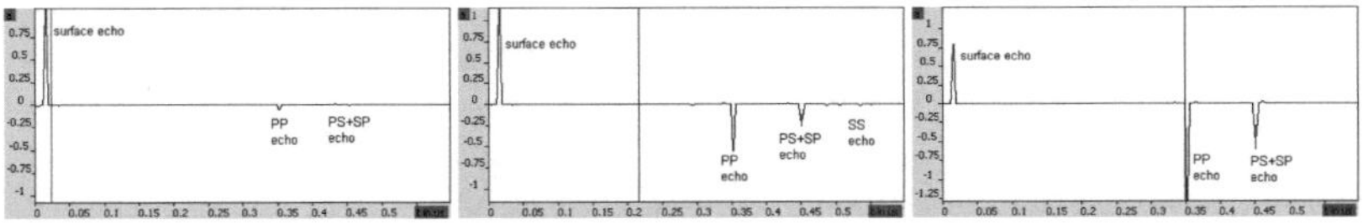

a) focus at surface **b) focus in center** **c) focus at backwall**

Figure 36: Profile multi-pulse deconvolution results from a 1 mm glass slide in water

5 Reconstruction

Reconstruction algorithms are used to solve inverse problems. The received electrical transducer signal is determined from the sound propagation in the specimen. The goal is now to determine information about the specimen, like structure and material characteristics, from the received signal. This can be done in reflective and/or transmission mode.

5.1 State of the art

For reflection mode tomography, projection theory can be used [60]. The reflection signals are recorded from all sides of the object and from this data the reflectivity of the object is reconstructed. Reference [60] also shows an approach for limited angle tomography. For transmission mode, ray tomography is convenient [61]. Only the first ray with minimum travel time is used, which can be determined by a combination of Fermat's principle and Huygens principle. Ultrasonic diffraction tomography also used in transmission mode [62]. Here, the whole scattered sound field at the receiver is used to determine the characteristics of the scattering objects.

The synthetic aperture focusing technique (SAFT) is designed to use the time signal at multiple positions around the specimen. The signals can be measured with an array or a scanning transducer. With a time and magnitude adjusted superposition of all signals, the aperture of the system can be adjusted to focus on any point within the object. The synthetic focusing of the excitation pulse is done with adjustable delay lines for the single elements of an array. It can be used in reflective and transmission mode. A reconstruction approach is demonstrated in reference [63]. The 3D object is modeled, and the SAFT reconstructed information from the real object is fed into the model. The resulting information can be displayed in a CAD environment. SAFT can produce different kinds of artifacts through mode converted waves and multiple reflections. They can be identified by comparison with a simulation [64].

In reference [65] the ultrasonic measurement is directly compared to simulations of all possible eventualities by pattern recognition on very simple objects. This approach can produce almost perfect results if only a small number of different situations are possible, and the simulations are representative. Up to now, it is almost impossible for complex structures due to the high quality requirements.

A ultrasonic microscope can also be used for material characterization. Phase velocities of different wave modes are easily determined if the sound path is known [66], [67]. For leaky waves it is very convenient to determine the phase velocity from interference patterns [68] with line focus transducers. The determination of density generally requires one or more calibration specimens. With spherically focused transducers, the reflection amplitude depends on a variety of factors and no direct measurement is possible. In the paper [69] a very promising method is introduced, which determines material parameters indirectly by comparing an ultrasonic simulation with the measured reflection signal. This approach is used for the characterization of materials of one single plane layer in water. It simplifies the system to a layered one-dimensional material structure and a sound field with rotatory symmetry and orthogonal incidence. All material parameters of water are well known, so the only unknowns are the material characteristics of the single layer.

5.2 Reconstruction approach

The reconstruction approach of paper [69] is not limited to one single layer. In this thesis it is extended to handle multiple layers. As long as there are at least two known materials, it is possible to characterize all other materials in the sound path. For a single reception signal system, the limitation is an one-dimensional layered material block. The direct material parameter estimation requires distinguishable reflections and a sufficient signal-to-noise ratio to determine all necessary reflection parameters. Most of this can be improved by applying the deconvolution to the received signal. More information and signal quality can be obtained by using multiple focus positions for the reconstruction process.

In the next stage, the directly estimated material parameters are used for an iterative optimization process, by comparing simulation results with the measurement. With each iteration step the precision of the single parameters is improved. Additionally, all parameters can be optimized for a maximum correlation between measured and simulated signal.

Finally, the complete 3D specimen can be classified point by point, and the three-dimensional material and structure information is obtained.

For simplification, the reconstruction equations are first derived for one isotropic material layer in water. Later, the reconstruction process is extended to handle multiple layers. Anisotropic materials can also be characterized by their average parameters in propagation direction. Single crystals are fully determined with separate characterizations for all main directions.

The stages for the complete reconstruction process are shown in figure 37. Not all stages are necessary for all applications. Depending on the demands and the information needed, only one or some work steps are required. In the following chapters each stage and different options are derived.

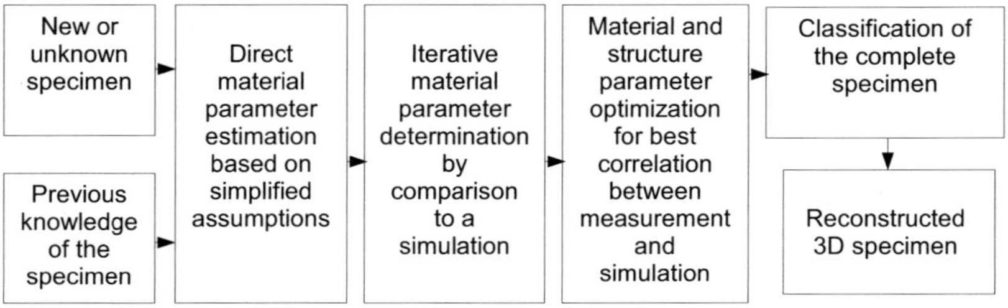

Figure 37: Reconstruction stages

5.3 Direct material parameter estimation

The direct estimation uses physical relations to determine material parameters directly from the reflection signal. Usually those relations have to be simplified with restrictive assumptions to solve the inverse problem analytically. The estimated results can also be used as initial values for the iterative material characterization. All equations are derived for one single layer in water and under the assumption of plane wave propagation.

5.3.1 Sound velocities and layer thickness

The sound velocity affects the time-of-flight through the one-layered specimen and the reflection amplitudes caused by material interfaces. Since the thickness of the specimen is usually known and time measurements are very precise, it is convenient to determine the velocity with simple propagation terms. The ray model (figure 24) leads to the following equations:

$$t_{\mathrm{PP}} - t_0 = \frac{2d}{c_{\mathrm{P}} \cdot \cos\theta_{\mathrm{P}}} \text{ and} \tag{43}$$

$$t_{\mathrm{SP}} - t_0 = \frac{d}{c_{\mathrm{P}} \cdot \cos\theta_{\mathrm{P}}} + \frac{d}{c_{\mathrm{S}} \cdot \cos\theta_{\mathrm{S}}}, \tag{44}$$

where d is the thickness of the specimen, t the echo time-of-flight, c the sound velocity, θ the angle of the wave normal and the indices P, S, W stand for the pressure (longitudinal), shear (transversal) and water wave. The index 0 represents the surface echo and multiple indices describe the wave path. With the assumption of plane waves, the relation between the wave angles is given by Snell's law:

$$\frac{c_W}{\sin\theta_W} = \frac{c_P}{\sin\theta_P} = \frac{c_S}{\sin\theta_S} \,. \tag{45}$$

The initial angles θ_W are determined from a sound field measurement (figure 10) if high precision results are required. For most applications of the ultrasonic microscope, the effective angles are lower than 5°. Therefore cos θ can be set to 1 for simplification and the sound velocities in equations (43) and (44) resolve now to:

$$c_P = \frac{2d}{t_{PP} - t_0} \text{ and} \tag{46}$$

$$c_S = \frac{d}{t_{SP} - t_0 - d/c_P} \,. \tag{47}$$

If the longitudinal sound velocity c_P is known in advance and the thickness d is unknown, d can be determined from equation (46). The highest precision for thickness determination is generally obtained by direct mechanical measurements. In many cases this involves high effort or requires destructive methods (cross-sections). With the methods presented here, ultrasonic microscopy can be used to non-destructively estimate the thickness of layers, even if they are buried.

Often c_P is unknown, whereas the sound velocity in water c_W is known very well. The time-of-flight in water over the distance of the specimen's thickness can be determined with a relative measurement to a comparative reflector as in figure 38. This method requires a water basin with a flat bottom, on which the specimen is placed. The difference of the time-of-flight from the bottom in water next to the specimen TOF_G and the surface time-of-flight TOF_{SF} from the surface of the specimen delivers the thickness d:

$$d = \tfrac{1}{2} \cdot c_W \cdot (TOF_G - TOF_{SF}) \tag{48}$$

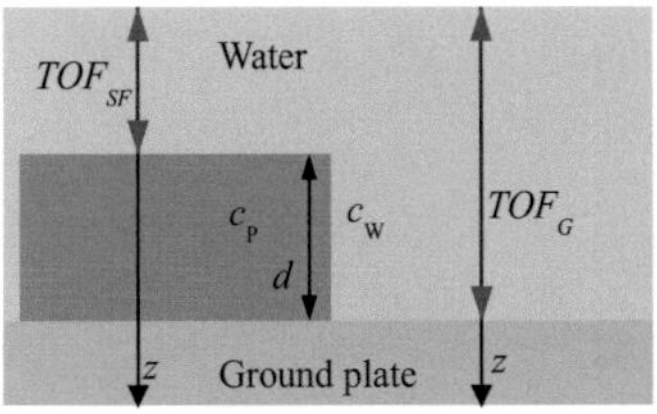

Figure 38: Relative thickness estimation

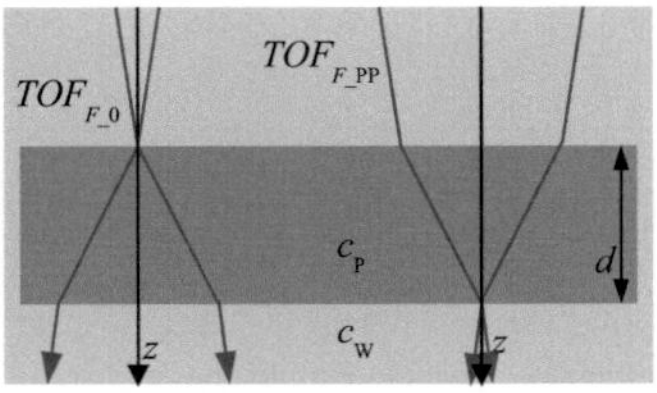

Figure 39: Thickness and sound velocity estimation with 2 focus positions

A new method is to estimate the thickness and the sound velocity with the generalized sound field position (*GSP*). With the *GSP* it is possible to predict the focus position in different material setups, and conversely, to determine the sound velocity and the thickness of one layer if the focus position is known. In focus, the *GSP* value of the surface echo (surface time-of-flight TOF_{F_0}) is the same as for the backwall echo in focus (surface time-of-flight TOF_{F_PP}, figure 39 and 55). The difference between surface echo and backwall echo can be determined with the sound velocity c_P:

$$c_W{}^2 \cdot TOF_{F\_0} = c_W{}^2 \cdot TOF_{F\_PP} + c_P{}^2 \cdot (t_{PP} - t_0). \tag{49}$$

For the thickness follows:

$$d = \tfrac{1}{2} \cdot c_W \cdot \sqrt{(TOF_{F\_0} - TOF_{F\_PP}) \cdot (t_{PP} - t_0)}, \tag{50}$$

and for the longitudinal sound velocity:

$$c_P = c_W \cdot \sqrt{\frac{TOF_{F\_0} - TOF_{F\_PP}}{t_{PP} - t_0}}. \tag{51}$$

This approach can also be used to estimate the shear wave sound velocity c_S. It does not require uncovered surfaces or single layers. Only the echoes have to be distinguishable. It is useful for buried layers where mechanical measurements or relative time-of-flight measurements are not possible. The results obtained with this approach are usually not very accurate and have a systematic discrepancy, which is caused by the simplified assumptions for the sound field modeling. With reference measurements and the corresponding correction factor for each transducer used, it provides very useful results as shown in table 4.

The estimation of sound velocity from the reflection amplitudes is also possible with equations (57) or (59), if the density is known. In most cases this is not recommended, since the amplitudes are affected by many other factors.

5.3.2 Density, elastic modules and acoustic attenuation

If the density ρ is known, the elastic modules can be determined from the sound velocities:

Young's modulus: $$E = 2 \cdot \rho \cdot c_S{}^2 \cdot \left(1 + \frac{c_P{}^2 - 2c_S{}^2}{2c_P{}^2 - 2c_S{}^2}\right) \tag{52}$$

Bulk modulus: $$K = \rho \cdot \left(c_P{}^2 - \frac{4}{3} c_S{}^2\right) \tag{53}$$

Poisson ratio: $$\upsilon = \frac{c_P{}^2 - 2c_S{}^2}{2c_P{}^2 - 2c_S{}^2} \tag{54}$$

It is also possible to determine the density from the elastic modules with the equations

(52) to (54), but the elastic modules are typically unknown.

The usual way for determining the density with ultrasound is the evaluation of the reflection amplitudes, which can be easily measured. They depend on a variety of factors like the focus position, attenuation, and reflection and transmission coefficients. A commonly used approach is the comparative estimation with a reference specimen. Although the surface reflection amplitude A_0 depends on a variety of factors, it is approximately proportional to the reflection coefficient for orthogonal incidence of plane waves at the same transducer distance. With this assumption, the density ρ can be estimated from the relationship of the surface reflection amplitude A_0 to a reference specimen's (density ρ_R, longitudinal sound velocity c_{RP}) surface reflection amplitude A_{R0} and the longitudinal sound velocity c_P with the same surface conditions:

$$\rho = \frac{\rho_W \cdot c_W \cdot (1+v)}{c_P \cdot (1-v)} \text{ with } v = \frac{A_0}{A_{R0}} \cdot \frac{\rho_R \cdot c_{RP} - \rho_W \cdot c_W}{\rho_R \cdot c_{RP} + \rho_W \cdot c_W}. \tag{55}$$

A more complex approach is introduced, which evaluates the amplitude of the surface echo and the longitudinal backwall echo A_{PP} from a plane layer inside water. Additionally ultrasonic attenuation a can be considered with the first longitudinal multiple reflection A_{PPPP}. Although the ultrasonic attenuation is frequency dependent, it can be modeled as a single factor for bandwidth limited ultrasonic reflections, as long as the dimensions are small and attenuation low. This attenuation factor a is characteristic for the material and the spectrum of the ultrasonic pulses. For simplification of the analytic solution, plane waves with orthogonal incidence are assumed for the determination of the reflection amplitudes:

$$A_0 = A \cdot \frac{Z - Z_W}{Z + Z_W}, \tag{56a}$$

$$A_{PP} = A \cdot \frac{4 \cdot Z \cdot Z_W \cdot (Z_W - Z)}{(Z + Z_W)^3} \cdot 10^{\frac{2 \cdot a \cdot d}{20\text{dB}}}, \tag{56b}$$

$$A_{PPPP} = A \cdot \frac{4 \cdot Z \cdot Z_W \cdot (Z_W - Z)^3}{(Z + Z_W)^5} \cdot 10^{\frac{4 \cdot a \cdot d}{20\text{dB}}}, \tag{56c}$$

where $Z = \rho c$ is the acoustic impedance and A the initial amplitude. The main advantages are that there is no reference or calibration required and the unknown parameters a and ρ are determined from one single reflection signal. The density from the equation system (56) is:

$$\rho = \frac{\rho_W c_W}{c_P A_{PP}^2}\left[A_{PP}^2 - 2A_0 A_{PPPP} + 2\,\text{sign}(A_0)\sqrt{A_0 A_{PPPP} \cdot \left(A_0 A_{PPPP} - A_{PP}^2\right)}\right], \tag{57}$$

and the attenuation is:

$$a = \frac{20\text{dB}}{2 \cdot d} \cdot \log_{10}\left(\frac{A_{\text{PPPP}}}{A_{\text{PP}}} - \frac{A_{\text{PP}}}{A_0}\right). \tag{58}$$

Without attenuation considerations ($a = 0$), the density estimation requires no multiple reflection:

$$\rho = \frac{\rho_\text{W} c_\text{W}}{c_\text{P} A_{\text{PP}}}\left[-A_{\text{PP}} - 2A_0 + 2\,\text{sign}\left(A_0\right)\sqrt{A_0 \cdot \left(A_{\text{PP}} + A_0\right)}\right]. \tag{59}$$

The sign(A_0) function is used to select the right solution depending on the material interface type (soft ↔ hard) to ensure that the initial amplitude is always positive ($A > 0$). The results are only estimates due to the one-dimensional approximation and correction methods should be applied.

A direct estimation of the acoustic attenuation can also be done by comparison of the backwall reflection amplitudes A_{PP} of two specimens of the same material but with different thickness d:

$$a = \frac{20\text{dB}}{2\left(d_2 - d_1\right)} \cdot \log_{10}\left(\frac{A_{\text{PP\_2}}}{A_{\text{PP\_1}}}\right). \tag{60}$$

5.4 Iterative material parameter determination of a single layer

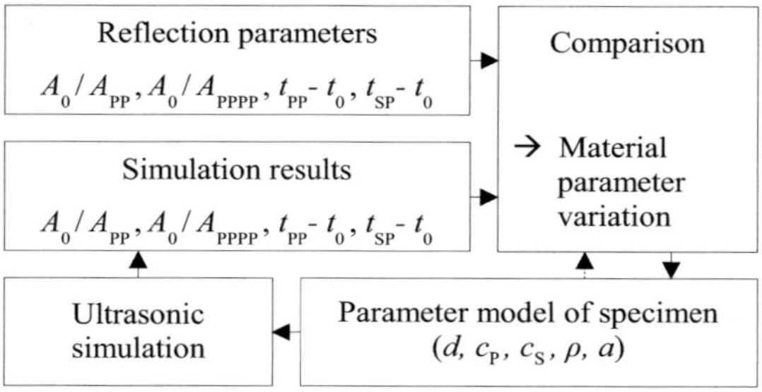

Figure 40: Iterative material characterization scheme

The iterative material characterization follows the inverse approach by comparing simulation results to the measured reflection parameters (figure 40). The initial values for the parameter model of the specimen are derived from the simplified direct equations (46) to (60), or they can be taken from a material database. Those initial parameters are used for the first material setup of the ultrasonic simulation. The resulting reflection signal from the simulation is analyzed the same way as the measured reflection signal, and the determined reflection parameters are compared. Depending on the complexity of the system (number of unknown material parameters), the parameter variation has to consider multiple variables and their dependence to each other, to determine a more accurate material parameter set. This new parameter set is used for the next iteration loop until a minimum error criterion is reached.

The resulting material parameters are as accurate as this criterion and the precision of the ultrasonic simulation. The total reconstruction time mainly depends on the simulation speed and the required iteration loops. One major aspect for this is the convergence speed of the algorithm used for the material parameter variation. Due to the possible complexity of the system, there might be more than one optimum if the system is under-determined only by the reflection parameters. For this reason, it is necessary to ensure that there are not more unknown material parameters than independent reflection parameters. One reflection signal from a single plane layer usually provides four independent usable parameters: the relation of the surface echo and the longitudinal backwall echo amplitude A_0/A_{PP}, the relation of the surface echo and the first longitudinal multiple reflection amplitude A_0/A_{PPPP}, the time difference between surface echo and backwall echo $t_{PP} - t_0$, and the time difference between surface echo and the first shear wave backwall echo $t_{SP} - t_0$. The basic relationship of those parameters with the material parameters is described by the equations (46) to (60). Since the number of material parameters for the simple case of one unknown material layer is already five, at least one of those parameters has to be known in order to be set constant in the reconstruction algorithm. This is usually the thickness, because it can be measured very precisely by other means. If more than one reflection signal is used at different focus positions, the additional information can be used to determine five unknown material parameters. The variation of the reflection amplitudes in dependence of the focus position provides information on sound velocities and thickness.

The parameter variation is done according to the error considerations, which are derived from the simplified direct equations. For one-dimensional sound propagation, they would lead to a very fast-converging and stable reconstruction system. Since the real system is much more complex, a variable loop gain factor G has to be applied to ensure the stability of the system and to provide optimum convergence. For simplicity, it can be set to a very low value, but this leads to a slow convergence. A better approach would be observation of the result and an automatic adjustment.

The new material parameters for each loop are obtained by the following equations. The new longitudinal sound velocity c_P derived from equation (46) is:

$$c_{P\_new} = c_P - G \cdot \frac{2 \cdot d}{\left(t_{PP} - t_0\right)^2} \cdot \left[\left(t_{PP\_ref} - t_{0\_ref}\right) - \left(t_{PP} - t_0\right)\right], \tag{61}$$

where the indices *_new* stand for the new value of the parameter and *_ref* stands for the reference value of the original reflection signal parameters. The new shear sound velocity based on equation (47) is:

$$c_{S\_new} = c_S - G \cdot \frac{d}{\left(t_{SP} - t_0 - d/c_P\right)^2} \cdot \left[\left(t_{SP\_ref} - t_{0\_ref}\right) - \left(t_{SP} - t_0\right)\right]. \tag{62}$$

The loop gain factor is particularly important for determining density and attenuation. The reflection amplitude is strongly influenced by other effects that have to be compensated for a stable reconstruction process. The new density according to equation (57) is:

$$\rho_{\_new} = \rho - G \cdot \frac{\rho_W \cdot c_W}{c_P} \cdot \left[2 - \text{sign}(A_0) \cdot \frac{2 \cdot v - 1}{\sqrt{v^2 - v}}\right] \cdot \left[v_{\_ref} - v\right], \text{ with } v = \frac{A_0^{\;2}}{A_{PP}^{\;2}} \cdot \frac{A_{PPPP}}{A_0} \tag{63}$$

and based on equation (59) without attenuation considerations (a=0):

$$\rho_{\_new} = \rho - G \cdot \frac{\rho_W \cdot c_W}{c_P} \left[2 - \text{sign}(A_0) \cdot \frac{2 \cdot v + 1}{\sqrt{v^2 + v}}\right] \cdot \left[v_{\_ref} - v\right], \text{ with } v = \frac{A_0}{A_{PP}}. \tag{64}$$

The new attenuation based on equation (58) is:

$$a_{\_new} = a + G \cdot \frac{20\text{dB} \cdot \left[\left(\frac{A_{PP\_ref}}{A_{0\_ref}} - \frac{A_{PP}}{A_0}\right)\left(-\frac{A_{PPPP} \cdot A_0}{A_{PP}^{\;2}} - 1\right) + \left(\frac{A_{PPPP\_ref}}{A_{0\_ref}} - \frac{A_{PPPP}}{A_0}\right) \cdot \frac{A_0}{A_{PP}}\right]}{2d \cdot \left(\frac{A_{PPPP}}{A_{PP}} - \frac{A_{PP}}{A_0}\right) \cdot \ln(10)}. \tag{65}$$

The loop gain G for each material parameter can be determined separately from the past result values: an oscillation requires a lower G value, and monotonous constant derivation is an indication for a too-low G value. For the ultrasonic simulation, it is convenient to use a fast and accurate algorithm. In general it is possible to use any type as long as it provides the reflection parameters. One goal of this thesis is high accuracy in an acceptable simulation time for practical use. The two-dimensional ray trace algorithm based on the sound field decomposition is very appropriate for this application and provides good results.

The convergence graph with a constant $G = 0.1$ for the characterization of fused silica (pure SiO_2) is shown in figure 41. Since it might happen that one variable becomes zero or below, a plausibility check and correction has to be performed after each iteration loop. The results of the reconstructed material parameters of fused silica are shown in table 1. They are also compared to the results from an ultrasonic finite element simulation (EFIT), with the same transducer and specimen (supplier data) as for the measurement. The EFIT simulation is without attenuation considerations. The thickness d was first measured mechanically (= supplier data) and then estimated with equation (48). Since the direct sound velocity estimation is very precise, an iterative determination was not performed. This leads to a lower precision for the shear wave velocity, because the effective shear wave angle is much higher than for the pure pressure wave (angle dependent mode conversion). Under laboratory conditions, the precision of the density determination with the ultrasonic microscope is relatively high.

Material parameter	ρ in kg/m³	d in µm	a in dB/m	c_P in m/s	c_S in m/s
Supplier data	2203	512	-	5965	3767
Direct estimation from SiO_2	2030	510	1547	6002	3630
Iterative determination from SiO_2	2181	-	27.9	-	-
Direct estimation from EFIT	1105	-	-	6006	3676
Iterative determination from EFIT	2178	-	-	-	-
Final relative measurement error	1%	<1%	-	<1%	3.6%

Table 1: Material parameters of a reconstructed SiO_2 layer; 75 MHz transducer

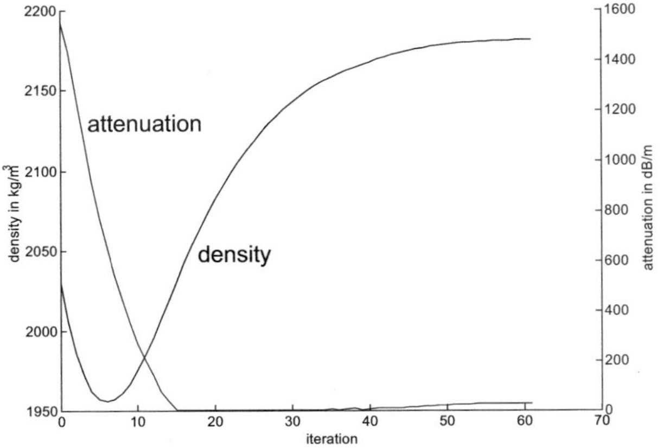

Figure 41: Convergence graph of the characterization of SiO_2; 75 MHz transducer Determination of ρ (black) and a (grey) with G = 0.1 and a cancellation criterion of 0.05% error

5.5 Reconstruction of complex specimens

In many cases, materials are assembled in a fixed structure where it is not possible to characterize the different materials independently from all others. The single layer reconstruction process is extended for more complex specimen composed of several unknown materials and/or with unknown structure. Generally, it is possible to characterize all materials of a layered structure, as long as all primary reflection parameters can be determined and at least two materials are known. The known materials can be the water of the delay line, the water behind the specimen, a known material inside the specimen or a secondary reference specimen for calibration. This reconstruction process requires at least the same number of independent input parameters as there are unknown variables in the whole system.

Based on the correlation comparison, an iterative reconstruction process for the local optimization of an unlimited number of material parameters from a known basis structure is developed. This approach needs no reflection parameter determination.

Another approach is developed based on the classification of unknown specimens with known simulated or measured reference structures.

5.5.1 Material characterization of multiple layers

The basic algorithm for the characterization of multiple layers is the same as for a single layer. All measured reflection parameters are used to determine the initial material parameters analytically from the one-dimensional simplified equations. Subsequently, the precision is improved with the iterative reconstruction process by comparison with the simulation results.

The thickness of each layer has to be known from mechanically measurements, or it can be determined with equation (50). The sound velocities are estimated with the same equations (46) and (47) as for one single layer. Depending on the number of unknown layers N, the analytic equations required for the density estimation lead to an equation system with N+1 unknowns. For one-dimensional wave propagation and with a water delay line the echo amplitudes correspond to:

$$A_0 = A \cdot r_0 \text{, with } r_0 = \frac{Z_0 - Z_W}{Z_0 + Z_W} \tag{66}$$

for the surface echo and without attenuation considerations

$$A_n = A \cdot r_n \cdot \prod_{i=0}^{n-1} (1 + r_i)(1 - r_i) \text{, with } r_i = \frac{Z_i - Z_{i-1}}{Z_i + Z_{i-1}} \tag{67}$$

for all following primary longitudinal interface echoes, where A is the initial amplitude, A_n the amplitude of the entry echo of layer n and $\rho = Z / c$ the density. At least one of the $N+1$ layers has to be known to determine A. With attenuation considerations, N additional equations are required that evaluate the first multiple reflection A_M of each layer:

$$A_{Mn} = -A \cdot r_{n-1} \cdot r_n^{\,2} \cdot 10^{\frac{2 \cdot a_{n-1} \cdot d_{n-1}}{20\mathrm{dB}}} \cdot \prod_{i=0}^{n-1} (1+r_i)(1-r_i) 10^{\frac{2 \cdot a_i \cdot d_i}{20\mathrm{dB}}} \tag{68}$$

and equation (67) is extended to:

$$A_n = A \cdot r_n \cdot \prod_{i=0}^{n-1} (1+r_i)(1-r_i) 10^{\frac{2 \cdot a_i \cdot d_i}{20\mathrm{dB}}} \,. \tag{69}$$

The parameter variation of the iterative reconstruction process can be done the same way as for a single layer. Equations (61) and (62) for the sound velocities remain valid. For density and attenuation, the single layer parameter variation with the equations (63) to (65) is only a local consideration. Dependencies of echoes from other layers are not considered, since the derivation of the exact error formulas is very complex and not necessarily better, due to the one-dimensional simplifications. The indices correspond as follows: $0 = n$, $\mathrm{PP} = n{+}1$ and $\mathrm{PPPP} = Mn{+}1$.

However, the convergence with local considerations is not very fast and there can be system oscillations depending on the individual gain factor G adjustment. The error formulas for the whole reconstruction system can also be determined numerically by slightly changing each input parameter for the simulation. This provides the derivations which are used for the error consideration formulas. For a system with a fixed number of layers the one-dimensional error terms can be derived analytically.

An example for the reconstruction of multiple layers is shown in figure 42. The cover foil materials of the first two layers of reference specimen 3 are unknown. Since only the primary longitudinal echoes are clearly detectable, the attenuation and shear wave velocity estimation is not possible. Common values are assumed here. The thicknesses have been determined mechanically and all other initial values are estimated with the one-dimensional analytical equations.

The convergence of the iterative multi-layer reconstruction is slower than for one layer, because it does not consider dependencies between the different layers. In the beginning phase of the reconstruction, the convergence speed is comparable to the single layer reconstruction. But in the later fine approximation phase it is much slower, as shown in figure 42. A dynamic variation of the gain value G might be used to improve this. The relative total error of the reflection amplitudes after 150 iterations with a constant $G = 0.05$ is about 0.05%.

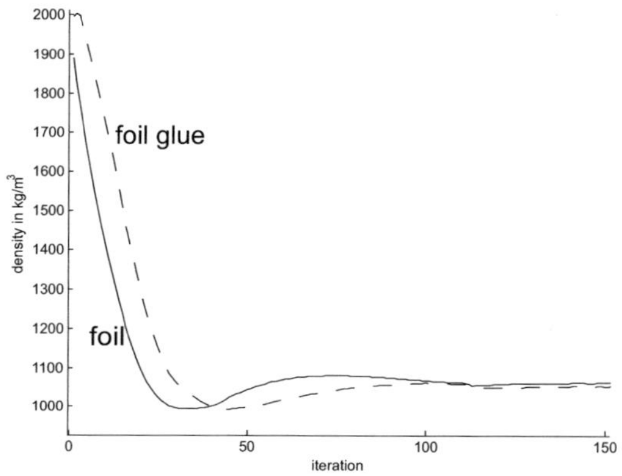

Figure 42: Convergence graph of the density estimation for the cover foil (solid) and the glue (dashed) of reference specimen 3; *G* = 0.05

5.5.2 Iterative simulation parameter optimization with correlation

In many cases it is not possible to separate the echoes from all interfaces. The layer-wise material characterization is one approach (see appendix page 95), which can help to obtain the reflection parameters from overlapping echoes. If the number of layers is too high, then the signal noise is stronger than the primary reflection amplitudes from deeper layers. In this case, it is not possible to perform a proper reconstruction.

Often the basic layer structure of a specimen is known. The first layers of such a specimen usually can be characterized with the above mentioned methods. For all deeper layers only the previous knowledge of the specimen can be used.

The idea of the iterative simulation parameter optimization is, to find the optimal simulation parameters, which deliver a simulation result as similar as possible to the measured data. For that, a simulation is set up with the basic knowledge of the specimen. The iterative simulation parameter optimization compares the result of the simulated specimen with the measured data with normalized cross-correlation of the ultrasonic signals. By slightly varying each simulation parameter P_i, the linearised error dependencies are determined for each parameter:

$$\frac{dCorr_i}{dP_i} = \frac{Corr(P_i + dP_i) - Corr(P_i - dP_i)}{2dP_i}, \tag{70}$$

where *Corr*(changed parameters) stands for the normalized cross-correlation result of the simulation signal with the measured signal, and *i* is the parameter index. The corresponding parameter is changed with the reduced error term:

$$P_{i\_new} = P_i + G \cdot \frac{1 - Corr}{dCorr_i} \cdot dP_i, \tag{71}$$

where *G* is the loop gain factor with a small positive value below one to ensure a stable reconstruction process. Additionally, plausibility checks and corrections are performed after each loop. For convenience, a parameter is only changed if the correlation is higher with the new parameter. Otherwise it is not changed. If the initial parameters are too far off from the actual parameters, this reconstruction process might find a local optimum and miss the global optimum. This can be avoided by determining all parameters as precisely as possible with other methods.

A convergence graph of the deterministic simulation parameter optimization for reference specimen 3 is shown in figure 43. The optimization was done for 15 material parameters and for the 4 different layer thicknesses. All layers of the same type are changed simultaneously. The most strongly varying parameters are usually the ultrasonic attenuations of the materials, because they especially affect the multiple reflections with low amplitudes. Such a precise evaluation of the multiple reflections is not possible with the methods above. Other parameters like sound velocities are only changed slightly, since they had been determined very precise earlier on.

The main advantage of this algorithm is the ability to use the information of the complete signal with all multiple reflections. Even if some of the primary echoes are below the signal noise, this algorithm can obtain the necessary information from the superimposing multiple reflection.

The optimization of 19 simulation parameters is a 19-dimensional problem. The deterministic search algorithm can only find the local optima, beginning with the initial parameter set. Statistical approaches can also find other and better local optima. For that, each parameter is changed with a Gaussian random process X, with a mean value of zero and the standard deviation σ:

$$P_{i\_new} = P_i \cdot [1 + \mathrm{X}(0, \sigma)]. \tag{72}$$

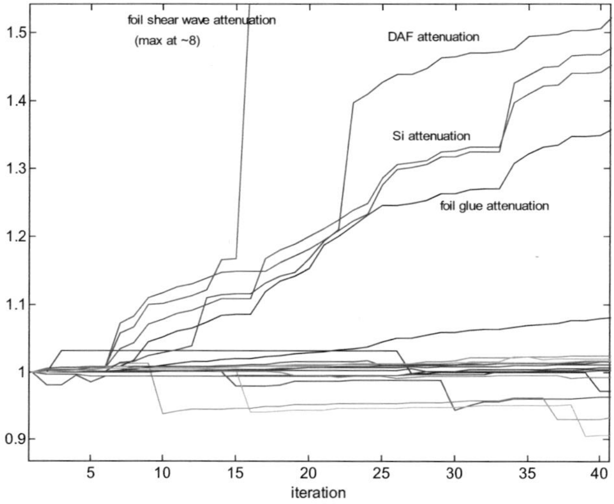

Figure 43: Convergence graph of the deterministic simulation parameter optimization relative changes; reference specimen 3; signal from position without defect; G = 0.05; dP_i / P_i = 0.0001; start correlation = 0.9310; correlation after 40 iterations = 0.9476

If the correlation with the new parameter is higher, then the new parameter is used. Otherwise the parameter is not changed. Plausibility checks and corrections are necessary after each iteration. The choice of σ influences the convergence of this algorithm. For a wide parameter search field and a good convergence, σ should be high (≈ 0.5) in the beginning, and then decrease with the number of iterations.

A convergence graph of the random simulation parameter optimization is shown in figure 44. In the beginning phase, the parameters vary widely, but stabilize with increasing numbers of iterations. The shear wave sound velocity parameter of the DAF tape is also changed substantially. It could not be measured before, so a guessed initial value was used here which is now optimized. The average correlation after 40 iterations is higher than with the deterministic search. The final material parameters after 40 iterations are listed in table 2. All parameters of the cover foil, the glue and the DAF tape are optimized. For the silicon only the attenuations are optimized, since all other parameters are well known. The active layer on the silicon surface is not considered which can diminish the accuracy of the determined density of the connected layers.

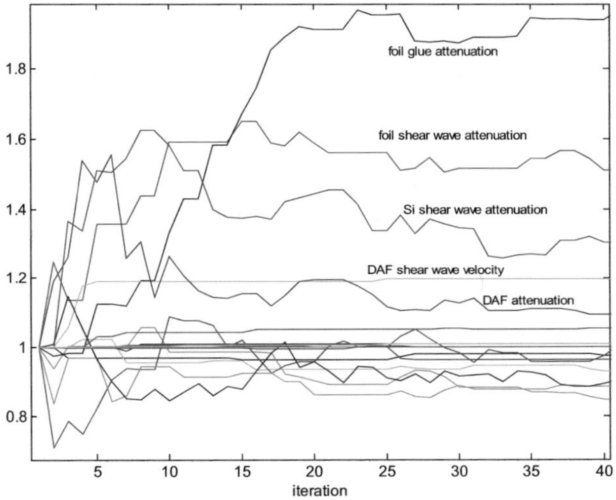

Figure 44: Convergence graph of the random simulation parameter optimization relative changes; reference specimen 3; signal from position without defect; average start correlation = 0.9336; correlation after 40 iterations = 0.9515

The waveforms of the optimized simulation and a measurement are compared in figure 45. It shows a high conformance for all strong echoes and fewer conformances of the weak multiple reflections. This is due to the correlation evaluation, where only the absolute difference is of interest. A relative error evaluation with a time-of-flight dependent variable gain control might improve the result, but this has not been tested yet.

Material Parameter	ρ in kg/m³	d in µm	c_P in m/s	c_S in m/s	a_P in dB/m	a_S in dB/m
Cover foil	1039	103.4	3107	1627	9953	11675
Cover foil glue	1355	11.0	1254	-	44042	-
Silicon	2330	73.9	8433	5843	17.1	23.8
DAF tape	1463	40.2	2230	881	7788	24475

Table 2: Determined material parameters of reference specimen 3 after 40 iterations

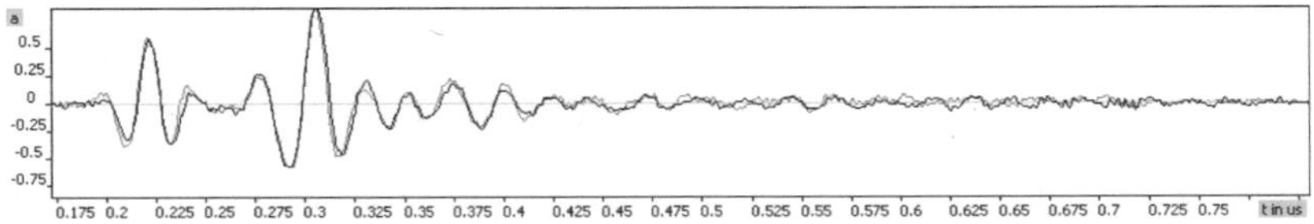

Figure 45: A-Scan comparison between simulation (grey) and measurement (black) reference specimen 3, position without defect; 75 MHz transducer

5.5.3 Pattern recognition reconstruction of specimens with known base structure

Pattern recognition can be used to classify an unknown specimen by comparing its characteristics with characteristics of a known specimen. If the characteristics of two specimens are very similar, then it is very likely that the two specimens are of the same kind. This classification approach can be used for the reconstruction of specimens with a known basic structure. Different types of flaws in the basic structure are simulated or can be measured on reference specimens. The results are compared with the measurements of the unknown specimen. According to the classification result, the three-dimensional structure information of the known specimen is assigned to the unknown specimen.

The major challenge of this approach is the choice of convenient feature vectors. For ultrasonic analysis, the best representative characteristics are the reflection parameters that are directly related to the investigated structure. They can be easily determined from the ultrasonic signal, if the echoes are non-interfering. On complex specimens it usually is not possible to obtain those reflection parameters for all primary echoes.

An efficient alternative is the direct comparison of the ultrasonic signals with cross-correlation. The one with the highest correlation most likely represents the measured structure. Due to the high sensitivity of complex systems, relatively strong changes in the signal can be caused by very small changes in the structure. Therefore, the correlation approach requires homogeneous structures or a huge knowledge database with all possible variations of the ultrasonic signal.

The reference specimen 3 has a complex and relatively homogeneous structure (see Appendix pages 97-99). All thicknesses of the layers have been determined with direct mechanically measurements. The material parameters for the first layers have been determined with the multiple layer reconstruction algorithm. Since all deeper layers are repeated, the same properties as in the first layers are assumed here. The optimization of the simulation parameters has been done with the iterative local optimization algorithm in chapter 5.5.2.

The goal of this classification is to find and to characterize flaws inside the specimen. For that, the classification signals for all possible defect cases have to be determined. In the case of reference specimen 3, delaminations and air inclusions between the single layers are of interest. In the reference simulations they are modeled with air as the replacement material for the corresponding layer behind the interface of interest. As measured reference, the same specimen with its known defect positions is used to measure the classification signals. Other specimens with the same structure, but without known defects, can be classified with the same classification signals.

Figure 46 shows the classification result with simulated classification signals. The displayed color represents the best matching classification signal and the brightness corresponds to the correlation value (table 3). Most of the known artificial air inclusions are classified correctly. There are some discrepancies at positions, where the specimen has a non-planar structure, like on the edges of the flaws (diffraction) and at the surface dent from the air inclusion of the first layer. Here, the signal correlation is generally much lower than at positions with almost perfect planar structure. More likely, a lower signal correlation can cause false classification results. In the case of the flaw at the 7th silicon layer, the correlation with the simulated signal with no defect is sometimes higher than with the simulated signal with the defect. The simulation quality is not sufficient here. Possible reasons are inaccurate simulation parameters (the active layer of the silicon is not modeled) and the discrepancy of the simulation model with reality (e.g. no frequency dependent attenuation).

Classification is also possible with measured classification reflection signals for the different defects. Such a result is shown in figure 47, and the corresponding correlation values are listed in table 3. As expected, the correlation between the signals is higher than with the simulated references. At a perfect planar structure, the only differences are caused by the random signal noise, which is about 3% of the total reflection signal energy. The comparison of the correlation results shows an average difference of about 2.5% between measured and simulated references. Thus, the ultrasonic simulation has an accuracy of about 97.5%.

Defect Si layer	none	1	2	3	4	5	6	7	8
Correlation with the simulated reference	0.951	0.828	0.929	0.941	0.949	0.947	0.954	0.948	0.951
Correlation with the measured reference	0.972	0.851	0.941	0.973	0.976	0.974	0.976	0.971	0.973

Table 3: Average normalized correlation of the reflection signals from reference specimen 3
The displayed value is the average correlation of the best matching classification signal at the correctly classified defect position

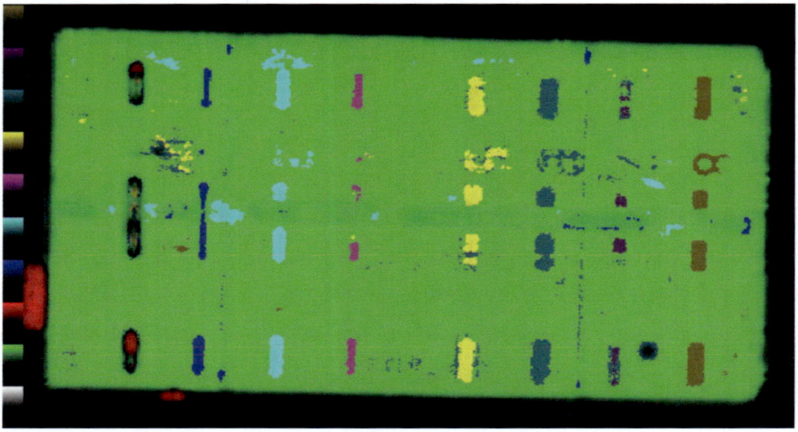

Figure 46: Classification C-Scan of reference specimen 3 with cross-correlation
scan area: 20.0 mm x 10.7 mm; simulated classification signals for the different known flaws; the color represents the best matching classification signal: white: unclassified, green: no delamination, red: air at the 1st silicon interface, blue: air at the 2nd silicon interface, cyan: air at the 3rd silicon interface, magenta: air at the 4th silicon interface, yellow: air at the 5th silicon interface, dark cyan: air at the 6th silicon interface, dark magenta: air at the 7th silicon interface, dark yellow: air at the 8th silicon interface

Figure 47: Classification C-Scan of reference specimen 3 with cross-correlation
scan area: 20.0 mm x 10.7 mm; manually selected classification signals from known flaw locations; colors as in figure 46

False classifications are also occurring with the measured references, especially at the edges of the flaws. At intermediate delaminations a proper classification is not possible, since there is no corresponding reference. At those positions the correlation value is generally lower.

The classification results can be used to perform a three-dimensional reconstruction of the specimen. Each class has an assigned planar structure, and this can be set together pixel by pixel according to the classification result. The obtained 3D volume of the specimen contains the material information at each position. This data can be displayed with 3D visualization software as is used for computer tomography.

In figure 48 the air inclusions of the reconstructed 3D volume are shown. All other materials are transparent. The reference air inclusions are located at their physical position, which makes a direct defect analysis and localization possible. A cross-sectional view is shown in figure 49. Compared to the cross-section B-Scan image (figure 1a) it shows reconstructed information, which is already interpreted and can provide the same information like physical cross-sections. A comparison with x-ray tomography is shown in figure 50. Due to the low absorption contrast between the materials, it is not possible to separate the different materials very clearly. Thin delaminations are not visible, because they have almost no volume to absorb x-rays. In contrast to the x-ray data, the reconstructed structure from the ultrasonic data also contains very thin delaminations, which are shown as air inclusions. There is no information about the layer structure behind a delamination, because the sound is totally reflected at an air interface.

Figure 48: 3D volume of the reconstructed reference specimen 3
transparent view – only air inclusions are visible (dark); based on classification results (figure 46)

Figure 49: 3D volume cross section of reference specimen 3
based on classification results (figure 46); the flaws of all 8 silicon layers are visible; blue: cover foil, red: DAF tape, grey: silicon, white: air, cyan: glass substrate

Figure 50: 3D volume cross section of reference specimen 3
X-ray computer tomography; the flaws of all 8 silicon layers are visible; red: mainly DAF tape, cover foil and air, grey: mainly silicon and glass; the contrast between air inclusions and DAF tape is very low

6 Applications and results

The algorithms developed in this thesis are all designed to improve the ultrasonic signal analysis for practical applications. For that, an ultrasonic analysis program called "USAnalyser" has been developed, which contains the advanced data analysis, the ultrasonic simulation and the reconstruction. Having everything in one program, all possible information can be obtained and be used by all algorithms to get more information. Some scientific relevant features of this program are demonstrated here.

6.1 Analysis of stacked components

The main purpose of this program is the analysis of stacked components with several dies. Since those industrial components usually have no properly known defects, the reference specimen 3 ("Ref3", appendix pages 97-99) has been created to evaluate and to demonstrate the developed algorithms.

With conventional analysis, only the defects of the first three dies could be visualized with sufficient signal-to-noise ratio. The focus is on the corresponding interface and the peak of the primary interface echo is displayed. By focusing deeper, the amplitude of the superimposed multiple reflections is increased, and all defects could be visualized as shown in figure 1a. The signal changes occur significantly later than the primary echo, which makes it difficult to determine the location and the kind of the defect. The desired information would be the material interfaces with a defect, as it can be obtained from figure 49 after the reconstruction.

Usually, the echoes of the first layers can be investigated directly by applying the deconvolution to the ultrasonic reflection signal (figure 32). As long as there are not too many closely spaced echoes, it provides clearly separated reflection information. On Ref3 the first three interfaces could be separated and displayed in a B-Scan in figure 59.

The investigation of all later echoes requires knowledge of how possible defects affect the reflection signal in order to determine their type and location. This can be done with a reference specimen with reference defects or with an ultrasonic simulation. Reference specimens are not always easy to make and they have to be exactly like the investigated component. On the other hand, ultrasonic simulations require the knowledge of all material parameters, which usually is not available. For that, the program has implemented several reconstruction algorithms, which make it easy to determine the material parameters (figure 37). All available information on the specimen can be used to determine all necessary material parameters.

As first step, the basic layer structure of the specimen is entered into the program. The material parameters used here are not required to be accurate, but at least for the layer thicknesses they should be. In the next step, the materials of the first layers are determined by selecting the primary echoes in the deconvoluted data. Internally, the program first estimates the material parameters with the simplified direct equations (46) to (60), and then iteratively determines the material parameter more and more accurately by applying changes according to the comparison of simulation and measurement. The obtained information for the first layers can be used for all deeper layers, since those are repeated periodically.

After the iterative material parameter determination, the simulated reflection signals already show a high similarity to the measured signals. Signal changes due to delamination at the different interfaces occur almost at the same time as in the measurement. For a direct comparison of the signals, the simulation parameters are optimized in the next step to make the simulated signals as similar as possible to the measured ones. This can be done with a deterministic local search (figure 43) or a random search (figure 44). The random searching algorithm usually can find better matches, since it is not bound to one local optimum. The final optimized material parameters of Ref3 are listed in table 2 and a comparison of the reflection signals is shown in figure 45.

For all delaminations at the different die interfaces, the corresponding simulation signals are highly correlated (table 3) and signal changes occur at the same times as in the measurement (figure 27a and 27c). In then next step, the measured signals are compared with the simulated signals. The highest correlated simulation signal most likely represents the defect type at the position of the measured signal. The classification image thus obtained, as in figure 46, directly shows the defect type and location for each pixel of the image. By using the structural information from the simulation, the three-dimensional structure of the specimen is obtained and can be displayed with a standard 3D visualization software as in figure 48 and 49. Compared to the original B-Scan in figure 1a, the defect type and location is directly visible.

Industrial components usually are encapsulated in molding compound, which can be very inhomogeneous (e.g. figure 51b). Here, the signals can not be compared directly at all positions with the same simulated structure. One possible approach is the material characterization of the molding compound for single pixels or small areas. As long as the signal-to-noise ratio is good enough, useful results can be obtained. Most of the structural noise comes from scattering in the molding compound. In figure 51 the analysis results of reference specimen 5 are shown, which is an encapsulated 5-die-stack component. Although the first die has no defects, there is a lot of structural noise visible in figure 51c, which is generated from the molding compound. All following dies have similar air inclusions as Ref3, and by analyzing the multiple reflections enough contrast is gained. Comparing the image of interest with the images of the earlier interfaces helps to find the differences, and the air inclusions can be identified. Even the substrate is very well visible and the shadows of all air inclusions in front appear.

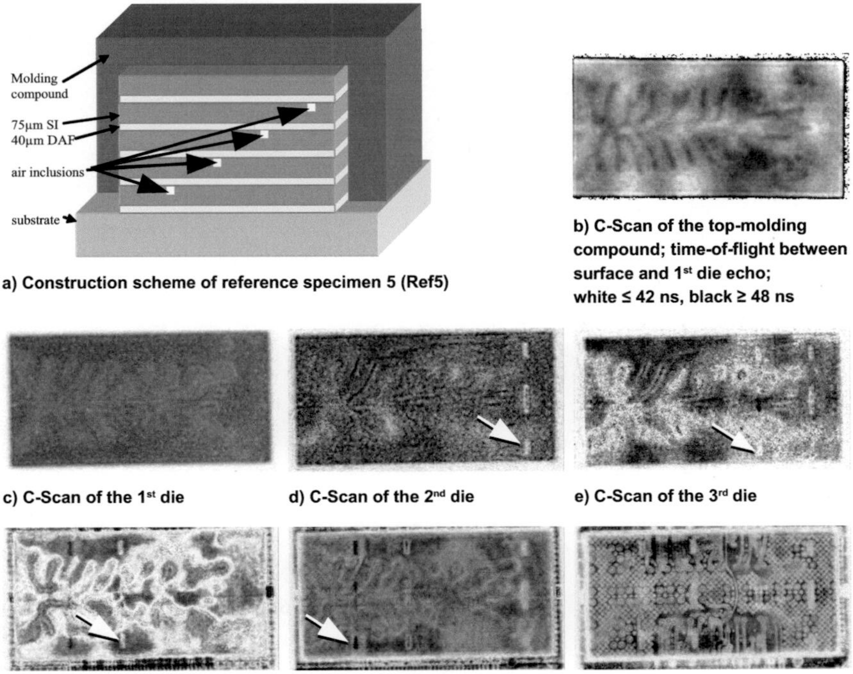

a) Construction scheme of reference specimen 5 (Ref5)

b) C-Scan of the top-molding compound; time-of-flight between surface and 1st die echo; white ≤ 42 ns, black ≥ 48 ns

c) C-Scan of the 1st die

d) C-Scan of the 2nd die

e) C-Scan of the 3rd die

f) C-Scan of the 4th die

g) C-Scan of the 5th die

h) C-Scan of the substrate

Figure 51: Analysis of reference specimen 5; five 18.5 mm x 9.5 mm dies stacked on a substrate and encapsulated with molding compound; all but the first die have an artificial air inclusion; amplitude of deconvoluted data; 75 MHz transducer; focus inside the substrate

The analysis of components with molding compound generally is not simple, since inhomogeneities of the filler distribution cause different sound velocities and thereby different interference situations for the multiple reflections. Those can be analyzed with time-of-flight measurements as in figure 51b. It is also possible to abrade the molding compound for higher quality investigations.

6.2 Time-of-flight and material analysis

There are many applications where the relatively high time resolution of the deconvolution can provide about ten times higher precision than conventional peak detection. By applying the deconvolution, the entire information of an ultrasonic pulse is basically averaged, the noise is reduced and the quality improved. On non-interfering pulses, the same precision can also be obtained by determining the time of the highest correlation between a previously selected pulse and the ultrasonic reflections. The precise maximum of the cross-correlation is found with sub-sample interpolation.

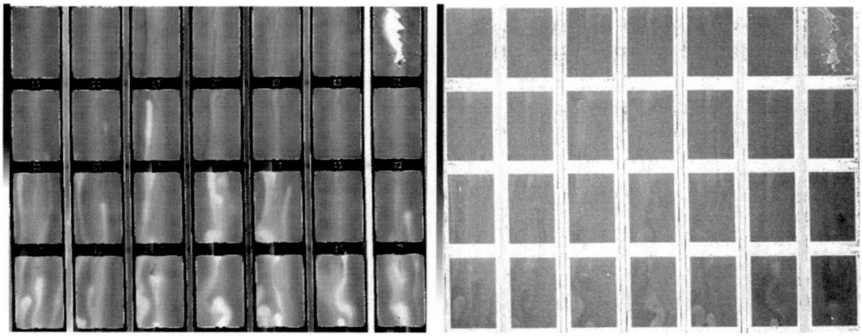

a) Time-of-flight analysis of the molding compound with deconvoluted data; time between surface echo and die echo: white ≤ 115ns, black ≥ 145ns **b) Amplitude analysis of the molding compound to die echo**

Figure 52: C-Scan of electronic components; scan area: 72 mm x 55 mm

In figure 52a the time-of-flight through the molding compound of electronic components is displayed. The distribution of the filler particles of the molding compound is seen very well, since it greatly affects the sound velocity. This information can be used to analyze and to improve the molding process. In combination with the amplitude information (figure 52b), the average elastic material characteristics at each pixel can be determined. The time-of-flight image is also useful for identifying artifacts, generated by the inhomogeneous molding compound, in the analysis of deeper interfaces (figure 51).

The sound velocity is not only sensitive to the material composition. It also depends on the stress situation in the specimen. This characteristic can be used for the detection of closed and open cracks, since those lead to a relaxation of the material. The crack in a silicon carbide layer and some large grain boundaries are easily seen in the time analysis image in figure 53a. One effect is relaxation, which leads to a slower sound velocity and thereby to a longer time (darker). Another effect is diffraction causing longer sound paths. In the amplitude image 53b only the open crack is well visible, since the amplitude is more sensitive to the material interfaces. The total dynamic range of the time (also called phase) image is only 9 ns at a sampling rate of 500 MHz. Due to the interpolation, a precision better than 0.1 ns for the relative times is achieved. There is no sampling or quantization noise visible, and the image has a very high quality.

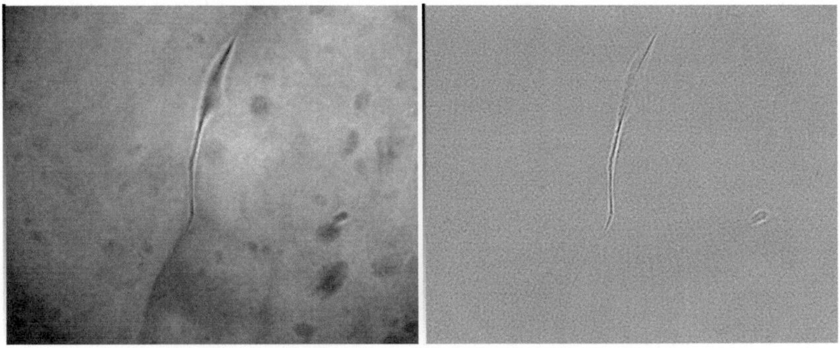

a) Time-of-flight analysis between surface and longitudinal backwall echo with cross-correlation; white ≤ 2075 ns, black ≥ 2084 ns

b) Amplitude analysis of the backwall shear-wave echo

Figure 53: C-Scans of a sintered 12.5 mm thick silicon-carbide layer partially open vertical crack visible; scan size 170 mm x 135 mm

A new application for time measurements is the simultaneous estimation of the sound velocity and the thickness of one layer. This layer can be completely unknown and buried inaccessible by mechanical means. The approach with two focus positions is described in chapter 5.3.1, and delivers the sound velocity and the thickness from the geometry of the sound field with equation (49).

For illustration, two sound field simulations have been performed, one with the focus point at the surface and one at the backwall. The maximum center normalized amplitudes of the simulation are displayed in figure 54. The difference of the surface echo time-of-flight is higher than the time between surface and backwall echo. Both times are used for the estimation. An example Z-Scan of a 700 µm glass slide is shown in figure 55. The times for the surface echo time-of-flight with the focus on the surface TOF_{F_0} and with the focus on the back TOF_{F_PP} are estimated from the maximum amplitude of the corresponding interface reflection. In reality, this is not exactly the focus position, but close to it.

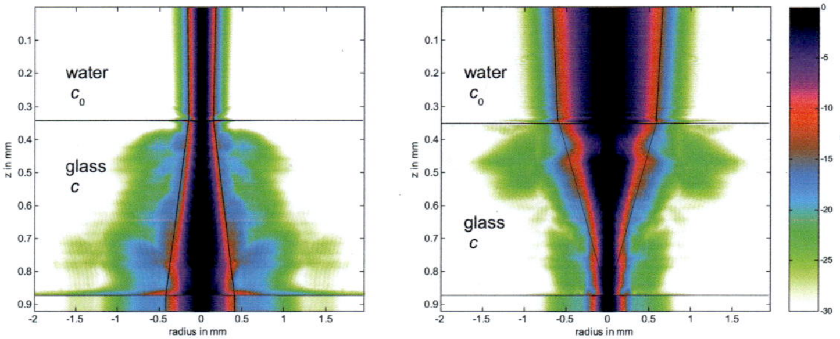

a) focus at surface (surface *TOF* = 36 µs) **b) focus at backwall (surface *TOF* = 33 µs)**

Figure 54: Sound field shape at a 512 µm SiO_2 glass slide; 75 MHz transducer
Sound field shape from an EFIT simulation; maximum center normalized amplitude

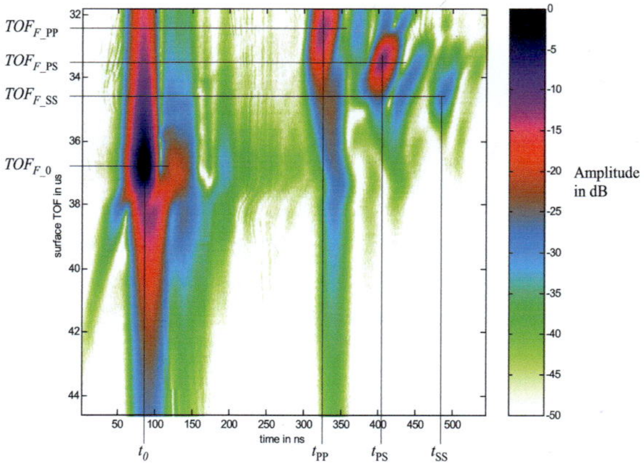

Figure 55: Z-Scan of a 700 µm borosilicate glass slide; 75 MHz transducer
Focus positions of the different reflection modes

Some results of the simultaneous estimation of thickness and sound velocity for several materials with different thicknesses are shown in table 4. The thicknesses *d* have been measured mechanically for comparison with the ultrasonic measurements. The evaluation of the times with equation (50) delivers a thickness that is higher than the actual thickness. This relative error increases with the sound velocity compared to water, and is caused by the idealized geometric assumptions for equation (49). Compensation is possible with a correction factor, which is calibrated on a known specimen with a similar sound velocity. This has been done with SiO_2 as reference specimen, and consequently the thickness for all glass based specimens is determined much more accurately. Other specimens like ceramic or plastic have a higher relative error. Without compensation the plastic specimen has the lowest error, since its sound velocity is closer to water.

Much better results can be obtained with a liner sound velocity depended correction factor $1 - k_d \cdot (c_P - c_W) / c_W$, where k_d is the thickness calibration constant from a known specimen and the used transducer. Applying this correction factor to equation (50) delivers a compensated thickness:

$$d = \frac{0.5 \cdot (1 + k_d) \cdot c_W \cdot \sqrt{(TOF_{F\_0} - TOF_{F\_PP}) \cdot (t_{PP} - t_0)}}{1 + k_d \cdot \sqrt{(TOF_{F\_0} - TOF_{F\_PP}) / (t_{PP} - t_0)}} . \qquad (73)$$

The results are now much more accurately with a final relative error in the range of 1%. Only for the ceramic specimen the error is relatively high, which is a hint to a nonlinear dependency of this systematic error and other dependencies like mass density. Future developments can be nonlinear compensation equations or to perform comparative sound field simulations and to determine the thickness and the sound velocities iteratively.

The same approaches are used for the determination of the sound velocity. In table 4 c_P is determined uncompensated from equation (51) and with the compensated thickness according to equation (46): $c_P = 2d / (t_{PP} - t_0)$. This approach delivers sound velocities and the thickness simultaneously, by evaluation of the time of flight and the focus positions.

Material mechanically measured d	SiO_2 512 µm	bs. glass 700 µm	glass 1100 µm	ceramic 1012 µm	plastic 2300 µm
$TOF_{F\_0} - TOF_{F\_PP}$	3272 ns	4176 ns	6771 ns	11927 ns	4885 ns
$t_{PP} - t_0$	171.9 ns	247.3 ns	382.2 ns	195.6 ns	1951 ns
geometric d from eq.(50)	563 µm	763 µm	1207 µm	1146 µm	2317 µm
geometric d from eq.(50) calibrated on 512 µm SiO_2	-	694 µm	1098 µm	1042 µm	2108 µm
geometric d from eq.(73) with correction factor $1 - 0.0304 \cdot (c_P - c_W) / c_W$	512 µm	699 µm	1103 µm	955 µm	2278 µm
geometric c_P from eq.(51)	6549 m/s	6168 m/s	6318 m/s	11721 m/s	2375 m/s
geometric c_P from eq.(73) and $c_P = 2d / (t_{PP} - t_0)$ with correction factor	5957 m/s	5653 m/s	5772 m/s	9765 m/s	2335 m/s

Table 4: Thickness and sound velocity estimation with different focus positions
75 MHz transducer; c_W = 1501 m/s (26.5°C), $_\Delta t$ = 1 ns, $_\Delta TOF$ = 50 ns, $_\Delta d$ = 2 µm, SiO_2: c_P = 5965 m/s, 96% Al_2O_3 ceramic, k_d = 0.0304 from SiO_2

7 Conclusions and perspectives

The purpose of this thesis was to develop advanced analysis tools for inspections with ultrasonic microscopes. This could be achieved though a comparative system analysis and an efficient system description.

An accurate ultrasonic simulation is essential for all iterative inverse reconstruction algorithms. For practical usage it should also be fast. The simulation algorithm developed here is based on the decomposition of the sound field into plane waves, and the plane wave propagation in plane layered structures is simulated with a ray-trace algorithm. Due to the high number of layers in the investigated specimens, the ray-trace algorithm has been optimized to reduce the mathematical effort. The exponential computational effort growth with the number of layers could be transformed into a linear dependency, which makes complex simulations possible in a very short simulation time. For the applications presented here, the quality and precision of this simulation algorithm is relatively high. There are several wave modes, which can not be considered, since the plane wave propagation approach does not consider diffraction and interface waves. In reality, those wave modes occur with some significance. The simulated pure longitudinal echoes are generally very precise, since they are not so much affected by much mode conversion.

The analysis of the ultrasonic signals and the reflection parameter estimation has been improved with a deconvolution algorithm. First, this deconvolution algorithm predicts the shape of the received pulses for a layered specimen in dependence of their time-of-flight. This can be done with previous knowledge of the specimen or with average assumptions. This information can also be obtained with a reconstruction process. With the knowledge of the corresponding reference signal for each time-of-flight, the deconvolution separates the echoes of received ultrasonic signals. A proper separation of two interfering echoes is possible down to a minimal distance of half a wavelength. Closer spaced echoes can not be separated independently.

Several inverse approaches for the reconstruction of the specimen under the ultrasonic microscope have been developed and tested. First, direct approaches with simplified analytical considerations are investigated. If the thickness of a single or multiple layers is known, very precise results for the sound velocity can be obtained, since relative time-of-flight measurements can be performed with a precision better than 1 ns. A new method is presented, which determines sound velocity and thickness of one or multiple layers at once –

without knowledge of the specimen and even of buried layers. This can be achieved with the additional analysis of the focus positions of the layer interfaces. The results with the new method can have a precision in the range of 99%. Physical values that can only be determined from the reflection amplitudes are generally not well determined, since the reflection amplitudes depend on multiple factors. Many of those factors, like the focus position, can be considered in an ultrasonic simulation. Because of that, the comparative iterative reconstruction processes can deliver much higher precision. Their quality only depends on the simulation accuracy. A relative error of about 1% for all elastic material characteristics can be achieved under laboratory conditions. The iterative reconstruction can be used for different analysis situations. It has been developed for specimens with multiple unknown layers. An iterative parameter optimization process can be used to minimize the discrepancies between simulation and measurement. The complete reconstruction process has been demonstrated on a very complex specimen with 19 layers, where a separation of all primary echoes is impossible.

The defect and structure classification, based on the correlation of the received signals with classification signals, delivers the three-dimensional structure of the specimen. The classification signals can be obtained from an ultrasonic simulation, or they can be measured on a reference specimen. After classification, the complete specimen can be displayed in a 3D visualization software. The results can be interpreted much more readily than the visualized ultrasonic signals. One major requirement for a successful classification is a relatively homogeneous specimen with a layered structure. Heavily scattering materials and/or structures can lead to false classifications.

The new methods and algorithms have also been demonstrated on real specimens. It was possible to find all artificial air inclusions on a realistic five die stack package with molding compound. Artifacts in the images from the molding compound inhomogeneities can be identified with a time-of-flight scan from the molding compound.

The high precision of the relative time-of-flight measurements enables high quality investigations on material homogeneity for analyzing material compositions, finding cracks and large grain boundaries.

The developed algorithms can also be used to improve the results of the classical ultrasonic analysis. With the deconvolution, amplitudes and times-of-flight can be determined more than 10 times more precisely than with classical peak detection. This reduces noise and provides better quality and ease of interpretation. The deconvolution also helps to separate superimposed reflections on thin layers and enables very stable pulse triggering and phase detection. With a simulation this close to reality, it is possible to perform feasibility investigations, to find an optimal measurement setup, and to understand and interpret results with more accuracy and confidence.

Perspectives

The investigations for an improved ultrasonic signal analysis showed several directions in which development could be continued. The ultrasonic simulation used here cannot handle several wave modes, which are caused by diffraction. Those wave types cannot be modeled with plane waves. For some high precision applications it might be advantageous to use more time consuming and more realistic simulation algorithms, which are based on point source synthesis.

Another aspect where simulation quality can be improved is frequency dependent attenuation. In this thesis, an average attenuation is considered for the center frequency of the pulse. This leads to signal shape discrepancies, if the simulated materials have a different frequency dependent attenuation than the water of the delay line during the reference signal acquisition. Spectral and shape changes of the pulses can be modeled with frequency dependent attenuation, as it is done for plane wave propagation in reference [17]. The compensation of the attenuation differences of the reference signals would also improve the quality of the signal deconvolution and cause fewer artifacts. The additional mathematical effort for frequency dependent attenuation is not very high, so the efficiency of the presented algorithms would be preserved.

The evaluation of ultrasonic signals from inhomogeneous scattering materials, like molding compound, poses more general challenges. It is not possible to perform a three-dimensional reconstruction of the scatterers from an one-dimensional ultrasonic signal. There are approaches to estimate the scatterer parameters from ultrasonic measurements, as in reference [70] and [71], but the results are not satisfying yet. It is possible to simulate wave propagation in scattering materials with finite elements methods, as in [72] and [12], to estimate mean parameters of statistical effects like damping and energy distribution. This can be helpful for understanding the effects on measured data. An exact modeling of the scattering effects would require a three-dimensional reconstruction process and data acquisition.

A possible approach is described in reference [63] with the synthetic aperture focusing technique (SAFT). Such approaches might also be applied to ultrasonic microscopes, since they can record the ultrasonic signals over a wide scanning area. Practical feasibility investigations and implementation could be a topic of future work. For the classical one-dimensional data evaluation, statistical approaches could be used to model the average variations due to scatterers over an area.

For all reconstruction approaches it would be helpful, if reflections from upper layers could be removed from the reflection signal, to reveal the reflections of deeper layers. Such an approach has been developed and is described in the appendix on the page 95. It has not been implemented jet.

Another issue for a better reconstruction process could be a better simulation parameter optimization process. The current one can only find local optima and thereby possibly miss the optimal global solution. Especially on highly complex specimens with many layers, there are numerous local optima very close to each other. It is possible to find the global optima by searching with a wide parameter field, but this would require a lot of processing time. Finding ways to reduce the parameter searching field would help to obtain better results faster.

With the new approach for the simultaneous estimation of the sound velocity and the thickness of one or more layers it should be possible to achieve a higher precision with an iterative method instead of the direct one. Comparing sound field simulations with the positions of the measured maximum amplitudes of the reflection signal would allow the compensation of systematic discrepancies and enable high quality measurements of sound velocities and thicknesses from completely unknown layers.

References

[1] : Jafar Saniie, Daniel T. Nagle, "Ultrasonic imaging though highly reverberant thin layers – theoretical considerations", IEEE Transactions on ultrasonics, ferrorelectrics and frequency control, vol 36, no 1, 1989, pp.80-92

[2] : Elfgard Kühnicke, "Elastische Wellen in geschichteten Festkörpersystemen", ISBN 3-934244-01-7, 2001, pp.

[3] : Martin spies, "Transducer field modeling in anisotropic media by superpositionof Gaussian base functions", Proceedings of the IEEE Ultrasonics Symposium, 1996, pp.685-688

[4] : Frank Schubert, Alexander Peiffer, Bernd, Terry Sanderson, "The elastodynamic finite integration technique for waves in cylindrical geometries", The Journal of the Acoustical Society of America, 1998, pp.2604-2614

[5] : E. Zhelezina, M. Kaltenbacher, R. Lerch, "Numerical simulation of acoustic wave propagation by a time and space adaptive finite element method", IEEE Ultrasonics symposium, 2002, pp.1213-1216

[6] : W. Lord, Z. You, M. Lusk, R. Ludwig, "Numerical predictions of surface wave phenomena for ultrasonic NDE", IEEE Ultrasonics Symposium, 1988, pp.1065-68

[7] : B.R. Tittmann, R. Lerch, H. Landes, S. Madden, H. Tavossi, "Wave propagation in long compound acoustic transmission line - numerical simulation and measurement", IEEE Ultrasonics symposium, 1997, pp.401-404

[8] : Sylvain Ballandras, Raphaël Lardat, Mikaël Wilm, Alexandre Reinhardt, Thomas Pastureaud, Vincent Laude, William Daniau, Raphaël Armati, William Steichen, Olivier Burat, "Simulation of 3D periodic piezoelectric transducers radiating in layered media using Finite Element/Boundary element Analysis", ECNDT 2006 - Fr.1.8.1, 2006, pp.

[9] : René Marklein, "Numerical Simulation of Fields and Waves in Nondestructive Testing", 2006 ECNDT Berlin - Tu.1.3.1, 2006, pp.1

[10] : P. Fellinger, R. Marklein, K. J. Langenberg, S. Klaholz, "Numerical modeling of elastic wave propagation and scattering with EFIT - Elastodynamic finite integration technique", Wave Motion 21, 1995, pp.47-66

[11] : Frank Schubert, Bernd Köhler, Alexander Peiffer, "Time Domain Modeling of Axisymmetric Wave Propagation in Isotropic Elastic Media With Cefit - Cylindrical Elastodynamic Finite Integration Technique", Journal of Computational Acoustics, 2001, pp.1127-1146

[12] : Frank Schubert, Bernd Köhler, "Three-dimensional time domain modeling of ultrasonic wave propagation in concrete in explicit consideration of aggregates and porosity", Journal of Computational Acoustics, vol 9, no 4, 2001, pp.1543-1560

[13] : Frank Schubert, René Marklein, "Numerical computation of ultrasonic wave propagation in concrete using the elastodynamic finite integration technique (EFIT)", IEEE Ultrasonics symposium, 2002, pp.799-804

[14] : Wagner Coelho, A. Pereira, David M. Simpson, Joiio C. Machado, "An estimator of focal position based on geometric acoustics", IEEE Ultrasonics Symposium, 1992, pp.323-325

[15] : Y. C. Chu, S. I. Rokhlin, "A method for determination of elastic constants of a unidirectional lamina from ultrasonic bulk velocity measurements on (0-90) cross-ply composites", Journal Acoustical Society of America, 1994, pp.342-352

[16] : Sébastien Deydier, Pierre Calmon, Odile Pétillon, "Modelling of the Ultrasonic Propagation into Carbon-Fiber-Reinforced Epoxy Composites Using a Ray-Theory-Based Homogenization Method", ECDNT Berlin, 2006, pp.1

[17] : Ping He, "Simulation of ultrasound pulse propagation in lossy media obeying a frequency power law", IEEE Transactions on ultrasonics, ferroelectrocs and frequency control, vol 45, no1, 1998, pp.114-125

[18] : Lars Odegaard, Sverre Holm, Frode Teigen, Trond Kleveland, "Acoustic field simulation for arbitrarily shaped transducers in a stratified medium", IEEE Ultrasonics symposium, 1994, pp.1535-1538

[19] : Pierre Calmon, "Simulation of Complex Configurations", ECNDT Berlin, 2006, pp.

[20] : Reinhard Lerch, Hermann Landes, Hans T. Kaarmann, "Finite element modeling of the pulse-echo behavior of ultrasound transducers", IEEE Ultrasonics symposium, 1994, pp.1021-1025

[21] : Stephane G. Mallat, Zhifeng Zhang, "Matching pursuits with time-frequency dictionaries", IEEE Transactions on signal processing vol 41, no 12, 1993, pp.3397-3415

[22] : Rémi Gribonval and Emmanuel Bacry, "Harmonic decomposition of audio signals with matching pursuit", IEEE Transactions on signal processing, vol 51, no1, 2003, pp.101-111

[23] : Günes, Z. Karabulut, Tolga Kurt, Abbas Yongacoglu, "Angle of arrival detection by matching pursuit algorithm", IEEE, 2004, pp.324-328

[24] : Ramazan Demirli, Jafar Saniie, "Denoising and compression of ultrasonic signals using model-based estimation techniques", IEEE International Ultrasonics, Ferroelectrics,and Frequency Control Joint 50th Anniversary Conference, 2004, pp.2306-2309

[25] : Sacha Krstulovic, Remi Gribonval, Pierre Leveau, Laurent Daudet, "A comparison of two extensions of the matching pursuit algorithm for the harmonic decomposition of sounds", IEEE Workshop on Applications of Signal Processing to Audio and Acoustics, 2005, pp.259-262

[26] : Alireza Shoa, Shahram Shirani, "Distortion of matching pursuit - modeling and optimization", IEEE Proceedings of the Data Compression Conference, 2006, pp.

[27] : Ramazan Demirli, Jafar Saniie, "A high-fidelity time-frequency representation for ultrasonic signal analysis", IEEE ULTRASONICS SYMPOSIUM, 2003, pp.1376-1379

[28] : R. Demirli, J. Saniie, "High resolution parameter estimation of ultrasonic echoes for NDE applications", IEEE Ultrasonics Symposium, 1997, pp.661-664

[29] : R. Demirli, J. Saniie, "Parameter estimation of multiple interfering echoes using the SAGE algorithm", IEEE Ultrasonics Symposium, 1998, pp.831-834

[30] : R. Demirli, J. Saniie, "Ultrasonic microscopy using low frequency transducers", IEEE Ultrasonics Symposium, 1999, pp.589-592

[31] : Ramazan Demirli, Jafar Saniie, "Model-based estimation of ultrasonic echoes. Part I - Analysis and algorithms", IEEE Transactions on ultrasonics, ferroelectrics, and frequency control, vol. 48, no. 3, 2001, pp.787-802

[32] : Ramazan Demirli, Jafar Saniie, "Model-based estimation of ultrasonic echoes. Part II - Nondestructive evaluation applications", IEEE Transactions on ultrasonics, ferroelectrics, and frequency control, vol. 48, no. 3, 2001, pp.803-811

[33] : Ramazan Demirli, Jafar Saniie, "Model based time-frequency estimation of ultrasonic echoes for NDE applications", IEEE Ultrasonics Sympositum, 2000, pp.785-788

[34] : J. Morlet, G. Arens, E. Fourgeau, D. Glard, "Wave propagation and sampling theory", Geophysics, Volume 47, 1982, pp.203-236

[35] : Ingrid Daubechies, "The wavelet transform, time-frequency localization and signal analysis", IEEE Transactions on Information theory, vol 36, no. 5, 1990, pp.961-1005

[36] : A. Abbate, J. Koay, J. Frankel, S.C. Schroeder, P. Das, "Application of wavelet transform signal processor to ultrasound", IEEE Ultrasonics Symposium, 1994, pp.1147-1152

[37] : Guilherme Cardoso, Jafar Saniie, "Data compression and noise suppression of ultrasonic NDE signals using wavelets", IEEE Ultrasonics Symposium, 2003, pp.250-253

[38] : Guilherme Cardoso, Jafar Saniie, "Ultrasonic data compression via parameter estimation", IEEE Transactions on ultrasonics, ferroelectrics, and frequency control, vol. 52, no. 2, 2005, pp.313-325

[39] : C. Sánchez-Ávila, "Wavelet domain signal deconvolution with singularity-preserving regularization", Mathematics and Computers in Simulation, 2003, pp.165–176

[40] : Jianqing Fan, Ja-Yong Koo, "Wavelet deconvolution", IEEE Transactions on Information Theory, vol. 48, no. 3, 2002, pp.734-747

[41] : Y. Lu, R. Demirli, G. Cardoso, J.Saniie, "Chirplet transform for ultrasonic signal analysis and NDE applications", IEEE Ultrasonics Symposium, 2005, pp.536-539

[42] : C.H. Chen, Wei-Lien Hsu, Sam-Kio Sin, "A comparison of wavelet deconvolution techniques for ultrasonic NDT", IEEE, 1988, pp.867-870

[43] : Sam-Kit Sin, Chi-Hau Chen, "A comparison of deconvolution techniques for the ultrasonic nondestructive evaluation of materials", IEEE Transactions on Image processing, 1992, pp.3-10

[44] : Abdelhak Bennia, Sedki M. Riad, "Filtering capabilities and convergence of the Van-Cittert deconvolution technique", IEEE Transactions on Instrumentation and measurement. volL. 41. no. 2., 1992, pp.246-250

[45] : Asoke K. Nandi, Detlef Mämpel, Burkhard Roscher, "Comparative study of deconvolution algorithms with applications in non-destructive testing", IEEE, 1995, pp.1-6

[46] : M.S. Unluturk, J. Saniie, "Deconvolution neural networks for ultrasonic testing", IEEE ULTRASONICS SYMPOSIUM, 1995, pp.715-719

[47] : Gordon Hayward, John E. Lewis, "A theoretical approach for inverse filter design in ultrasonic applications", IEEE Transactions on Ultrasonics, Ferroelectrics and Frequency control, vol. 36, no 3, 1989, pp.356-364

[48] : Abdelhak Bennia, Norris S. Nahman, "Deconvolution of causal pulse and transient data", IEEE Transactions on Instrumentation and Measurement, vol 79, no 6, 1990, pp.933-939

[49] : Edward J. Rothwell, Weimin Sun, "Time domain deconvolution of transient radar data", IEEE Transactions on Antennas and Propagation, vol 38, no 4, 1990, pp.470-475

[50] : T. Dhaene, L. Martens, D. De Zutter, "Generalized iterative frequency domain deconvolution technique", IEEE, 1993, pp.85-87

[51] : Tom Dhaene, Luc Martens, Daniel De Zutter, "Extended Bennia-Riad criterion for iterative frequency-domain deconvolution", IEEE Transaction on Instrumentation and Measurements, vol. 42, no. 2, 1994, pp.176-180

[52] : Ramesh Neelamani, Hyeokho Choi, Richard Baraniuk, "ForWaRD - Fourier-wavelet regularized deconvolution for ill-conditioned systems", IEEE Transactions on signal processing, vol 52, no 2, 2004, pp.418-433

[53] : Richard Gut, Peter Kreiert, George S. Moschytz, "Trellis-based deconvolution of ultrasonic echoes", IEEE, 1995, pp.1384-1387

[54] : Y. Goussard, G. Demoment, J. Idier, "A new algorithm for iterative deconvolution of sparse spike trains", IEEE, 1990, pp.1547-1550

[55] : F. Champagnat, J. Idier, G. Demoment, "Deconvolution of sparse spike trains accounting for wavelet phase shifts and colored noise", IEEE, 1993, pp.452-455

[56] : Qiansheng Cheng, Rong Chen, Ta-Hsin Li, "Simultaneous wavelet estimation and deconvolution of reflection seismic signals", IEEE Transactions on geoscience and remote

sensing, vol 34, no 2, 1996, pp.377-384

[57] : Frederic Champagnat, Yves Goussard, Jerome Idier, "Unsupervised deconvolution of sparse spike trains using stochastic approximation", IEEE Transactions on signal processing, vol 44, no 12, 1996, pp.2988-2998

[58] : J. Martinsson, F. Hägglund, J.E. Carlson, "Complete post-separation of overlapping ultrasonic signalsby combining hard and soft modeling", Ultrasonics 48, 2008, pp.427–443

[59] : Norbert Gust, "Untersuchung zur Verbessung der Signalauswertung beim Ultraschallmikroskop", Diplomarbeit TU-Dresden, 2005, pp.19-24

[60] : C. Q. Lan, W. Xiong, "An iterative method of ultrasonic reflection mode tomography", IEEE Transactions on medical imaging, vol 13, no2, 1994, pp.419-425

[61] : Lin-ping Song, Shu-yi Zhang, "Stabilizing the iterative solution to ultrasonic ray tomography", IEEE Ultrasonics symposium, 1997, pp.1497-1500

[62] : Xiaochuan Pan, Mark A. Anastasio, Chien-Min Kao, "A comparative study of image reconstructions in SPECT and ultrasonic diffraction tomography", IEEE Transactions on nuclear science, vol 46, no 3, 1999, pp.527-534

[63] : A. McNab, D. Reilly, A. Potts, M. Toft, "Role of 3-D graphics in NDT data processing", IEEE Proc.-Sei. Meas. Technol., Vol. 148, No. 4, 2001, pp.149-158

[64] : V. Schmitz, K.J. Langenberg, S. Chakhlov, "Calculation of high frequency ultrasonic signals for shear wave insonification in solid material", Ultrasonics 42, 2004, pp.249–252

[65] : L. Peter Martin, David H. Chambers, Graham H. Thomas, "Experimental and simulated ultrasonic characterization of complex damage in fused silica", IEEE Transactions on ultrasonics, ferrorelectronics and frequency control, vol 49, no 2, 2002, pp.255-265

[66] : N. Kim, "Simultaneous measurement of longitudinal and transverse waves velocities employing spherically focused acoustic transducer", Key Engineering Materials Volume 321-323 I (2006-01-01), Jan. 2006, pp.574-577

[67] : Christophe Graciet and Bernard Hosten, "Simultaneous measurement of speed, attenuation, thickness and density with reflected ultrasonic waves in plates", Ultrasonics symposium, 1994, pp.1219-1222

[68] : Jun-Ichi Kushibiki and Noriyoshi Chubachi, "Material Characterization by Line-Focus-Beam Acoustic Microscope", IEEE Transactions on sonics and ultrasonics, vol. SU-32, no. 2, March 1985, pp.189-212

[69] : Norbert Gust, Elfgard Kühnicke, Dirk Breuer, "Material characterization with the ultrasonic microscope", IEEE conference proceedings ISSE 2008 in Budapest, 2008, pp.92-95

[70] : R. Leo Romijn, Johan M. Thijssen, Gerald W. J. van Beuningen, "Estimation of scatterer size from backscattered ultrasound - a simulation study", IEEE Transactions on ultrasonics, ferroelectrics and frequency control, vol 36, no 6, 1989, pp.593-606

[71] : Claudio Simon, Jian Shen, Ralf Seip, Emad S. Ebbini, "A robust and computationally efficient algorithm for mean scatterer spacing estimation", IEEE Transactions on ultrasonics, ferrorelectrocs and frequency control, vol 44, no 4, 1997, pp.882-894

[72] : Frank Schubert, Bernd Koehler, "Numerical time-domain simulation of diffusive ultrasound in concrete", Ultrasonics 42, 2004, pp.781–786

Figures

Figure 1: Reference specimen 3: eight-die stack with air-inclusions....3
Figure 2: Conventional ultrasonic analysis scheme....4
Figure 3: Ultrasonic system analysis scheme....5
Figure 4: Sound field measurement scheme....9
Figure 5: Measurement of the self-convoluted sound field; 75 MHz transducer; scale in dB; 1000 times averaged data....10
Figure 6: Self-convoluted sound field after noise filtering; 75 MHz transducer; logarithmic scale in dB....10
Figure 7: Sound field (self-deconvolution of measurement); 75 MHz transducer; logarithmic scale in dB....10
Figure 8: Sound field at a constant z-level after noise filtering; 75 MHz transducer; logarithmic scale in dB....10
Figure 9: Wave front shape; 75 MHz transducer....11
Figure 10: Amplitude distribution in dependence of the propagation direction from equation (4); 75 MHz transducer....11
Figure 11: Reflection and transmission for pressure-wave incidence....11
Figure 12: Reflection and transmission for shear-wave incidence....11
Figure 13: Sound field decomposition scheme....14
Figure 14: Sound field propagation simulation based on the sound field measurement (figure 8)....17
Figure 15: Sound field propagation simulation based on the sound field measurement (figure 8)....18
Figure 16: Idealized sound field description with ray geometry at a planar material interface....19
Figure 17: Geometrical considerations at a spherical transducer....23
Figure 18: Amplitude curve – maximum amplitude in dependence of the TOF (transducer z-position)....23
Figure 19: Reference signal in dependence of surface time-of-flight TOF (focus position);....26
Figure 20: Comparison of a measured signal (red) and a calculated signal from a sound field measurement (black)....29
Figure 21: 1D ray trace model....31
Figure 22: Impulse response according to figure 21....31
Figure 23: A-Scan comparison of a 515 um SiO2 layer in water....33
Figure 24: 2D ray trace model....33
Figure 25: Reflection amplitudes from a single 1 mm glass slide....34
Figure 26: A-Scan comparison of a 515 µm SiO2 layer in water....34
Figure 27: Comparison of measured and simulated B-Scans from reference specimen 3....37
Figure 28: Z-Scan of a 1 mm glass slide....43
Figure 29: Absolute correlation of the surface echo with itself in dependence of the TOF (z-position of the transducer)....43
Figure 30: Absolute correlation between surface echo and longitudinal backwall echo PP....43
Figure 31: Iterative deconvolution scheme....45
Figure 32: A-Scan from reference specimen 3 (no flaw position); 75 MHz transducer....46
Figure 33: Deconvolution results from a 1 mm glass slide in water; 75 MHz transducer....47
Figure 34: GSP profiles for a 1 mm glass slide in water....48
Figure 35: Homogeneous multi-pulse deconvolution results from a 1 mm glass slide in water....48
Figure 36: Profile multi-pulse deconvolution results from a 1 mm glass slide in water....49
Figure 37: Reconstruction stages....52
Figure 38: Relative thickness estimation....53
Figure 39: Thickness and sound velocity estimation with 2 focus positions....53
Figure 40: Iterative material characterization scheme....56
Figure 41: Convergence graph of the characterization of SiO2; 75 MHz transducer....59
Figure 42: Convergence graph of the density estimation for the cover foil (solid) and the glue (dashed) of reference specimen 3; G = 0.05....62
Figure 43: Convergence graph of the deterministic simulation parameter optimization....64

Figure 44: Convergence graph of the random simulation parameter optimization....65
Figure 45: A-Scan comparison between simulation (red) and measurement (black)....65
Figure 46: Classification C-Scan of reference specimen 3 with cross-correlation....68
Figure 47: Classification C-Scan of reference specimen 3 with cross-correlation....68
Figure 48: 3D volume of the reconstructed reference specimen 3....70
Figure 49: 3D volume cross section of reference specimen 3....70
Figure 50: 3D volume cross section of reference specimen 3....70
Figure 51: Analysis of reference specimen 5; five 18.5 mm x 9.5 mm dies stacked on a substrate and encapsulated with molding compound; all but the first die have an artificial air inclusion; amplitude of deconvoluted data; 75 MHz transducer; focus inside the substrate....73
Figure 52: C-Scan of electronic components; scan area: 72 mm x 55 mm....74
Figure 53: C-Scans of a sintered 12.5 mm thick silicon-carbide layer....75
Figure 54: Sound field shape at a 512 µm SiO_2 glass slide; 75 MHz transducer....76
Figure 55: Z-Scan of a 700 µm borosilicate glass slide; 75 MHz transducer....76
Figure 56: Construction scheme of reference specimen 3 “ref3”....97
Figure 57: Silicon active side LASER air inclusion before cleaning....98
Figure 58: Overview C-Scan of Ref3 with B-Scan line....98
Figure 59: B-Scan of Ref3 through the flaws of all layers....98
Figure 60: C-Scans of reference specimen 3 from deconvoluted data....99

Tables

Table 1: Material parameters of a reconstructed SiO2 layer; 75 MHz transducer....59
Table 2: Determined material parameters of reference specimen 3 after 40 iterations....65
Table 3: Average normalized correlation of the reflection signals from reference specimen 3....67
Table 4: Thickness and sound velocity estimation with different focus positions....77

Appendix

Appendix 1: Derivation of the reflection and transmission coefficients....................................90
Appendix 2: Layer-wise reconstruction algorithm..95
Appendix 3: Reference specimen 3..97

Appendix 1: Derivation of the reflection and transmission coefficients

See figure 11 and 12 for the definitions of variables and coordinates. The reflection and transmission coefficients are determined for plane waves, which can be tilted in the *x*-*z*-plane and are independent of the *y* axis. The materials are restricted to be isotropic and the material interface (material 1 to material 2) is plane (independent of the *x* and *y* axis).

The boundary conditions at the material interface for the stress (tension) are:

$$\sigma_{xx1}=\sigma_{xx2},\sigma_{zz1}=\sigma_{zz2},\sigma_{xz1}=\sigma_{xz2},\sigma_{yy1}=0,\ \sigma_{yy2}=0,\ \sigma_{xy1}=0,\sigma_{xy2}=0,\sigma_{yz1}=0,\sigma_{yz2}=0 \quad (74a)$$

and for the strain:

$$\varepsilon_{x1}=\varepsilon_{x2},\ \varepsilon_{z1}=\varepsilon_{z2},\ \varepsilon_{y1}=0,\ \varepsilon_{y2}=0\,. \quad (74b)$$

The Hooke's law for an isotropic elastic medium is:

$$\bar{\sigma}=\bar{\bar{C}}\bar{\varepsilon} \quad (75)$$

with

$$\bar{\sigma}=\begin{pmatrix}\sigma_{xx}\\ \sigma_{yy}\\ \sigma_{zz}\\ \sigma_{xy}\\ \sigma_{xz}\\ \sigma_{yz}\end{pmatrix},\ \bar{\varepsilon}=\begin{pmatrix}\varepsilon_{xx}\\ \varepsilon_{yy}\\ \varepsilon_{zz}\\ \varepsilon_{xy}\\ \varepsilon_{xz}\\ \varepsilon_{yz}\end{pmatrix},\ \bar{\bar{C}}=\begin{pmatrix}\lambda+2\mu & \lambda & \lambda & 0 & 0 & 0\\ \lambda & \lambda+2\mu & \lambda & 0 & 0 & 0\\ \lambda & \lambda & \lambda+2\mu & 0 & 0 & 0\\ 0 & 0 & 0 & \mu & 0 & 0\\ 0 & 0 & 0 & 0 & \mu & 0\\ 0 & 0 & 0 & 0 & 0 & \mu\end{pmatrix},$$

where λ and μ are the Lamé constants with $\lambda=\rho\cdot(c_{\mathrm{P}}^2-2c_{\mathrm{S}}^2)$ and $\mu=\rho c_{\mathrm{S}}^2$.

With the boundary conditions (74), equation (75) is simplified to:

$$\sigma_{xx}=\left(\lambda+2\mu\right)\varepsilon_{xx}+\lambda\varepsilon_{zz} \quad (76a)$$

$$\sigma_{zz}=\left(\lambda+2\mu\right)\varepsilon_{zz}+\lambda\varepsilon_{xx} \quad (76b)$$

$$\sigma_{xz}=\mu\varepsilon_{xz} \quad (76c)$$

The solutions of the wave equation (5) for the sound particle velocity *v* and plane waves modes are listed below. Here *A* is the amplitude, *d* the oscillation direction, *k* the wave number, *ω* the angular frequency and *t* the time:

Incoming plane pressure wave:

$$\vec{v}_{0\mathrm{P}}=A_{0\mathrm{P}}\vec{d}_{0\mathrm{P}}\exp\left[j\cdot\left(\vec{k}_{0\mathrm{P}}\vec{r}+\omega t\right)\right] \text{ with } \vec{k}_{0\mathrm{P}}=\begin{pmatrix}k_{0\mathrm{P}x}\\ k_{0\mathrm{P}z}\end{pmatrix}=\frac{\omega}{c_{\mathrm{P1}}}\begin{pmatrix}\sin\theta_{\mathrm{P1}}\\ \cos\theta_{\mathrm{P1}}\end{pmatrix},\ \vec{d}_{0\mathrm{P}}=\begin{pmatrix}\sin\theta_{\mathrm{P1}}\\ \cos\theta_{\mathrm{P1}}\end{pmatrix}. \quad (77)$$

Incoming plane shear wave:

$$\vec{v}_{0S} = A_{0S}\vec{d}_{0S}\exp\left[j\cdot\left(\vec{k}_{0S}\vec{r}+\omega t\right)\right] \text{ with } \vec{k}_{0S} = \begin{pmatrix} k_{0Sx} \\ k_{0Sz} \end{pmatrix} = \frac{\omega}{c_{S1}}\begin{pmatrix} \sin\theta_{S1} \\ \cos\theta_{S1} \end{pmatrix},\ \vec{d}_{0S} = \begin{pmatrix} \cos\theta_{S1} \\ -\sin\theta_{S1} \end{pmatrix}. \quad (78)$$

Reflected plane pressure wave:

$$\vec{v}_{RP} = A_{RP}\vec{d}_{RP}\exp\left[j\cdot\left(\vec{k}_{RP}\vec{r}+\omega t\right)\right] \quad \text{with} \quad \vec{k}_{RP} = \begin{pmatrix} k_{RPx} \\ k_{RPz} \end{pmatrix} = \frac{\omega}{c_{P1}}\begin{pmatrix} \sin\theta_{P1} \\ -\cos\theta_{P1} \end{pmatrix},\quad \vec{d}_{RP} = \begin{pmatrix} \sin\theta_{P1} \\ -\cos\theta_{P1} \end{pmatrix}. \quad (79)$$

Reflected plane shear wave:

$$\vec{v}_{RS} = A_{RS}\vec{d}_{RS}\exp\left[j\cdot\left(\vec{k}_{RS}\vec{r}+\omega t\right)\right] \text{ with } \vec{k}_{RS} = \begin{pmatrix} k_{RSx} \\ k_{RSz} \end{pmatrix} = \frac{\omega}{c_{S1}}\begin{pmatrix} \sin\theta_{S1} \\ -\cos\theta_{S1} \end{pmatrix},\ \vec{d}_{RS} = \begin{pmatrix} \cos\theta_{S1} \\ \sin\theta_{S1} \end{pmatrix}. \quad (80)$$

Transmitted plane pressure wave:

$$\vec{v}_{TP} = A_{TP}\vec{d}_{TP}\exp\left[j\cdot\left(\vec{k}_{TP}\vec{r}+\omega t\right)\right] \text{ with } \vec{k}_{TP} = \begin{pmatrix} k_{TPx} \\ k_{TPz} \end{pmatrix} = \frac{\omega}{c_{P2}}\begin{pmatrix} \sin\theta_{P2} \\ \cos\theta_{P2} \end{pmatrix},\ \vec{d}_{TP} = \begin{pmatrix} \sin\theta_{P2} \\ \cos\theta_{P2} \end{pmatrix}. \quad (81)$$

Transmitted plane shear wave:

$$\vec{v}_{TS} = A_{TS}\vec{d}_{TS}\exp\left[j\cdot\left(\vec{k}_{TS}\vec{r}+\omega t\right)\right] \text{ with } \vec{k}_{TS} = \begin{pmatrix} k_{TSx} \\ k_{TSz} \end{pmatrix} = \frac{\omega}{c_{S2}}\begin{pmatrix} \sin\theta_{S2} \\ \cos\theta_{S2} \end{pmatrix},\ \vec{d}_{TS} = \begin{pmatrix} \cos\theta_{S2} \\ -\sin\theta_{S2} \end{pmatrix}. \quad (82)$$

The angles are related according to Snell's law:

$$\xi = \frac{\sin\theta_{P1}}{c_{P1}} = \frac{\sin\theta_{P2}}{c_{P2}} = \frac{\sin\theta_{S1}}{c_{S1}} = \frac{\sin\theta_{S2}}{c_{S2}}. \quad (83)$$

By placing the wave modes (77) to (82) into the boundary conditions (74) we obtain the following equations for an incoming pressure wave:

$$\sigma_{zz1} = \sigma_{zz0P} + \sigma_{zzRP} + \sigma_{zzRS} = \sigma_{zz2} = \sigma_{zzTP} + \sigma_{zzTS}\,|_{z=0}:$$

$$\begin{aligned}
&\sigma_{zz}(x,t)\,|_{z=0} = \rho_1 c_{P1} A_{0P}\exp\left[j\omega\cdot(\xi x+t)\right]\cdot\left((1-c_{P1}^2\xi^2)+(c_{P1}^2-2c_{S1}^2)\xi^2\right)+\\
&\rho_1 c_{P1} A_{RP}\exp\left[j\omega\cdot(\xi x+t)\right]\cdot\left((1-c_{P1}^2\xi^2)+(c_{P1}^2-2c_{S1}^2)\xi^2\right)-\\
&\rho_1 c_{S1} A_{RS}\exp\left[j\omega\cdot(\xi x+t)\right]\cdot\xi\sqrt{1-c_{S1}^2\xi^2}\,2 = \rho_2 c_{P2} A_{TP}\exp\left[j\omega\cdot(\xi x+t)\right]\cdot\\
&\left((1-c_{P2}^2\xi^2)+(c_{P2}^2-2c_{S2}^2)\xi^2\right)+\rho_2 c_{S2} A_{TS}\exp\left[j\omega\cdot(\xi x+t)\right]\cdot\xi\sqrt{1-c_{S2}^2\xi^2}\,(-2)
\end{aligned}$$

$$\rightarrow \begin{aligned}
&\rho_1 c_{P1}\cdot\left(A_{0P}+A_{RP}\right)\left(1-2c_{S1}^2\xi^2\right)-2\rho_1 c_{S1}^2 A_{RS}\xi\sqrt{1-c_{S1}^2\xi^2} = \\
&\rho_2 c_{P2} A_{TP}\cdot\left(1-2c_{S2}^2\xi^2\right)-2\rho_2 c_{S2}^2 A_{TS}\xi\sqrt{1-c_{S2}^2\xi^2}
\end{aligned} \quad (84)$$

$$\sigma_{xz1} = \sigma_{xz0P} + \sigma_{xzRP} + \sigma_{xzRS} = \sigma_{xz2} = \sigma_{xzTP} + \sigma_{xzTS}\,|_{z=0}:$$

$$\sigma_{xz}(x,t)|_{z=0} = j\omega\rho_1 c_{S1}^2 A_{0P}\exp\left[j\omega\cdot(\xi x+t)\right]\cdot\left(2\xi\sqrt{1-c_{P1}^2\xi^2}\right)+$$
$$j\omega\rho_1 c_{S1}^2 A_{RP}\exp\left[j\omega\cdot(\xi x+t)\right]\cdot\left(-2\xi\sqrt{1-c_{P1}^2\xi^2}\right)+j\omega\rho_1 c_{S1} A_{RS}\exp\left[j\omega\cdot(\xi x+t)\right]\cdot\left((1-c_{S1}^2\xi^2)-c_{S1}^2\xi^2\right)$$
$$=j\omega\rho_2 c_{S2}^2 A_{TP}\exp\left[j\omega\cdot(\xi x+t)\right]\cdot\left(2\xi\sqrt{1-c_{P2}^2\xi^2}\right)+j\omega\rho_2 c_{S2} A_{TS}\exp\left[j\omega\cdot(\xi x+t)\right]\cdot\left((1-c_{S2}^2\xi^2)-c_{S2}^2\xi^2\right)$$

$$\rightarrow \begin{array}{l}\rho_1 c_{S1}^2\cdot\left(A_{0P}-A_{RP}\right)\left(2\xi\sqrt{1-c_{P1}^2\xi^2}\right)+\rho_1 c_{S1} A_{RS}\cdot\left(1-2c_{S1}^2\xi^2\right)=\\ \rho_2 c_{S2}^2 A_{TP}\cdot\left(2\xi\sqrt{1-c_{P2}^2\xi^2}\right)+\rho_2 c_{S2} A_{TS}\cdot\left(1-2c_{S2}^2\xi^2\right)\end{array} \tag{85}$$

$$v_{x1}=v_{x0P}+v_{xRP}+v_{xRS}=v_{x2}=v_{xTP}+v_{xTS}|_{z=0}:$$

$$v_x(x,t)|_{z=0}=A_0 c_{P1}\xi\exp\left[j\omega\cdot(\xi x+t)\right]+A_{RP}c_{P1}\xi\exp\left[j\omega\cdot(\xi x+t)\right]+A_{RS}\sqrt{1-c_{S1}^2\xi^2}\cdot$$
$$\exp\left[j\omega\cdot(\xi x+t)\right]=A_{TP}c_{P2}\xi\exp\left[j\omega\cdot(\xi x+t)\right]+A_{TS}\sqrt{1-c_{S2}^2\xi^2}\exp\left[j\omega\cdot(\xi x+t)\right]$$

$$\rightarrow\quad c_{P1}\xi\cdot\left(A_0+A_{RP}\right)+A_{RS}\sqrt{1-c_{S1}^2\xi^2}=A_{TP}c_{P2}\xi+A_{TS}\sqrt{1-c_{S2}^2\xi^2} \tag{86}$$

$$v_{z1}=v_{z0P}+v_{zRP}+v_{zRS}=v_{z2}=v_{zTP}+v_{zTS}|_{z=0}:$$

$$v_z(x,t)|_{z=0}=A_0\sqrt{1-c_{P1}^2\xi^2}\exp\left[j\omega\cdot(\xi x+t)\right]-A_{RP}\sqrt{1-c_{P1}^2\xi^2}\exp\left[j\omega\cdot(\xi x+t)\right]+$$
$$A_{RS}c_{S1}\xi\cdot\exp\left[j\omega\cdot(\xi x+t)\right]=A_{TP}\sqrt{1-c_{P2}^2\xi^2}\exp\left[j\omega\cdot(\xi x+t)\right]-A_{TS}c_{S2}\xi\cdot\exp\left[j\omega\cdot(\xi x+t)\right]$$

$$\rightarrow\quad \left(A_0-A_{RP}\right)\sqrt{1-c_{P1}^2\xi^2}+A_{RS}c_{S1}\xi=A_{TP}\sqrt{1-c_{P2}^2\xi^2}-A_{TS}c_{S2}\xi \tag{87}$$

By replacing the amplitudes A with the reflection R and transmission T coefficients and 1 for the incident wave in equations (84) to (87), we obtain:

$$R_{PP}\rho_1 c_{P1}q_{S1}-R_{PS}2\rho_1 c_{S1}^2\xi w_{S1}-T_{PP}\rho_2 c_{P2}q_{S2}+T_{PS}2\rho_2 c_{S2}^2\xi w_{S2}=-\rho_1 c_{P1}q_{S1} \tag{88a}$$

$$-R_{PP}2\rho_1 c_{S1}^2\xi w_{P1}-R_{PS}\rho_1 c_{S1}q_{S1}-T_{PP}2\rho_2 c_{S2}^2\xi w_{P2}-T_{PS}\rho_2 c_{S2}q_{S2}=-2\rho_1 c_{S1}^2\xi w_{P1} \tag{88b}$$

$$R_{PP}c_{P1}\xi+R_{PS}w_{S1}-T_{PP}c_{P2}\xi-T_{PS}w_{S2}=-c_{P1}\xi \tag{88c}$$

$$-R_{PP}w_{P1}+R_{PS}c_{S1}\xi-T_{PP}w_{P2}+T_{PS}c_{S2}\xi=-w_{P1} \tag{88d}$$

with $q_{ab}=\left(1-2c_{ab}^2\xi^2\right)$ and $w_{ab}=\sqrt{1-c_{ab}^2\xi^2}$, where the index ab can be P1, P2, S1 or S2.

The equation system (88) can also be written in matrix form:

$$\begin{pmatrix} \rho_1 c_{P1} q_{S1} & -2\rho_1 c_{S1}^2 \xi w_{S1} & -\rho_2 c_{P2} q_{S2} & 2\rho_2 c_{S2}^2 \xi w_{S2} \\ -2\rho_1 c_{S1}^2 \xi w_{P1} & -\rho_1 c_{S1} q_{S1} & -2\rho_2 c_{S2}^2 \xi w_{P2} & -\rho_2 c_{S2} q_{S2} \\ c_{P1}\xi & w_{S1} & -c_{P2}\xi & -w_{S2} \\ -w_{P1} & c_{S1}\xi & -w_{P2} & c_{S2}\xi \end{pmatrix} \begin{pmatrix} R_{PP} \\ R_{PS} \\ T_{PP} \\ T_{PS} \end{pmatrix} = \begin{pmatrix} -\rho_1 c_{P1} q_{S1} \\ -2\rho_1 c_{S1}^2 \xi w_{P1} \\ -c_{P1}\xi \\ -w_{P1} \end{pmatrix} \tag{88}$$

For an incoming shear wave we obtain the following equations:

$$\sigma_{zz1} = \sigma_{zz0S} + \sigma_{zzRP} + \sigma_{zzRS} = \sigma_{zz2} = \sigma_{zzTP} + \sigma_{zzTS} \,|_{z=0}:$$

$$\begin{aligned} &\sigma_{zz}(x,t)\,|_{z=0} = \rho_1 c_{S1} A_{0S} \exp\left[j\omega\cdot(\xi x + t)\right]\cdot \xi\sqrt{1-c_{S1}^2\xi^2}\,(-2) + \\ &\rho_1 c_{P1} A_{RP} \exp\left[j\omega\cdot(\xi x + t)\right]\cdot\left((1-c_{P1}^2\xi^2) + (c_{P1}^2 - 2c_{S1}^2)\xi^2\right) - \\ &\rho_1 c_{S1} A_{RS} \exp\left[j\omega\cdot(\xi x + t)\right]\cdot \xi\sqrt{1-c_{S1}^2\xi^2}\,2 = \\ &\rho_2 c_{P2} A_{TP} \exp\left[j\omega\cdot(\xi x + t)\right]\cdot\left((1-c_{P2}^2\xi^2) + (c_{P2}^2 - 2c_{S2}^2)\xi^2\right) + \\ &\rho_2 c_{S2} A_{TS} \exp\left[j\omega\cdot(\xi x + t)\right]\cdot \xi\sqrt{1-c_{S2}^2\xi^2}\,(-2) \end{aligned}$$

$$\begin{aligned} \rightarrow\; &-2\rho_1 c_{S1}^2 A_{0S}\xi\sqrt{1-c_{S1}^2\xi^2} + \rho_1 c_{P1} A_{RP}\cdot\left(1-2c_{S1}^2\xi^2\right) - 2\rho_1 c_{S1}^2 A_{RS}\xi\sqrt{1-c_{S1}^2\xi^2} = \\ &\rho_2 c_{P2} A_{TP}\cdot\left(1-2c_{S2}^2\xi^2\right) - 2\rho_2 c_{S2}^2 A_{TS}\xi\sqrt{1-c_{S2}^2\xi^2} \end{aligned} \tag{89}$$

$$\sigma_{xz1} = \sigma_{xz0S} + \sigma_{xzRP} + \sigma_{xzRS} = \sigma_{xz2} = \sigma_{xzTP} + \sigma_{xzTS} \,|_{z=0}:$$

$$\begin{aligned} &\sigma_{xz}(x,t)\,|_{z=0} = j\omega\rho_1 c_{S1} A_{0S} \exp\left[j\omega(\xi x + t)\right]\cdot\left((1-c_{S1}^2\xi^2) - c_{S1}^2\xi^2\right) + \\ &j\omega\rho_1 c_{S1}^2 A_{RP} \exp\left[j\omega\cdot(\xi x + t)\right]\cdot\left(-2\xi\sqrt{1-c_{P1}^2\xi^2}\right) - j\omega\rho_1 c_{S1} A_{RS} \exp\left[j\omega\cdot(\xi x + t)\right]\cdot\left((1-c_{S1}^2\xi^2) - c_{S1}^2\xi^2\right) \\ &= j\omega\rho_2 c_{S2}^2 A_{TP} \exp\left[j\omega\cdot(\xi x + t)\right]\cdot\left(2\xi\sqrt{1-c_{P2}^2\xi^2}\right) + j\omega\rho_2 c_{S2} A_{TS} \exp\left[j\omega\cdot(\xi x + t)\right]\cdot\left((1-c_{S2}^2\xi^2) - c_{S2}^2\xi^2\right) \end{aligned}$$

$$\begin{aligned} \rightarrow\; &\rho_1 c_{S1} A_{0S}\cdot\left(1-2c_{S1}^2\xi^2\right) - \rho_1 c_{S1}^2 A_{RP}\cdot\left(2\xi\sqrt{1-c_{P1}^2\xi^2}\right) - \rho_1 c_{S1} A_{RS}\cdot\left(1-2c_{S1}^2\xi^2\right) = \\ &\rho_2 c_{S2}^2 A_{TP}\cdot\left(2\xi\sqrt{1-c_{P2}^2\xi^2}\right) + \rho_2 c_{S2} A_{TS}\cdot\left(1-2c_{S2}^2\xi^2\right) \end{aligned} \tag{90}$$

$$v_{x1} = v_{x0S} + v_{xRP} + v_{xRS} = v_{x2} = v_{xTP} + v_{xTS} \,|_{z=0}:$$

$$\begin{aligned} &v_x(x,t)\,|_{z=0} = A_{0S}\sqrt{1-c_{S1}^2\xi^2}\exp j\omega(\xi x + t) + A_{RP} c_{P1}\xi \exp j\omega(\xi x + t) + \\ &A_{RS}\sqrt{1-c_{S1}^2\xi^2}\exp j\omega(\xi x + t) = A_{TP} c_{P2}\xi \exp j\omega(\xi x + t) + A_{TS}\sqrt{1-c_{S2}^2\xi^2}\exp j\omega(\xi x + t) \end{aligned}$$

$$\rightarrow\; A_{0S}\sqrt{1-c_{S1}^2\xi^2} + A_{RP} c_{P1}\xi + A_{RS}\sqrt{1-c_{S1}^2\xi^2} = A_{TP} c_{P2}\xi + A_{TS}\sqrt{1-c_{S2}^2\xi^2} \tag{91}$$

$$v_{z1} = v_{z0S} + v_{zRP} + v_{zRS} = v_{z2} = v_{zTP} + v_{zTS}\,|_{z=0}:$$

$$v_z(x,t)\,|_{z=0} = -A_{0S}c_{S1}\xi \exp[j\omega\cdot(\xi x+t)] - A_{RP}\sqrt{1-c_{P1}^2\xi^2}\exp[j\omega\cdot(\xi x+t)] +$$
$$A_{RS}c_{S1}\xi \exp[j\omega\cdot(\xi x+t)] = A_{TP}\sqrt{1-c_{P2}^2\xi^2}\exp[j\omega\cdot(\xi x+t)] - A_{TS}c_{S2}\xi \exp[j\omega\cdot(\xi x+t)]$$

$$\rightarrow\ -A_{0S}c_{S1}\xi - A_{RP}\sqrt{1-c_{P1}^2\xi^2} + A_{RS}c_{S1}\xi = A_{TP}\sqrt{1-c_{P2}^2\xi^2} - A_{TS}c_{S2}\xi \tag{92}$$

By replacing the amplitudes A with the reflection R and transmission T coefficients and 1 for the incident wave in equations (89) to (92), we obtain:

$$R_{SP}\rho_1 c_{P1} q_{S1} - R_{SS}2\rho_1 c_{S1}^2\xi w_{S1} - T_{SP}\rho_2 c_{P2} q_{S2} + T_{SS}2\rho_2 c_{S2}^2\xi w_{S2} = 2\rho_1 c_{S1}^2\xi w_{S1} \tag{93a}$$

$$-R_{SP}2\rho_1 c_{S1}^2\xi w_{P1} - R_{SS}\rho_1 c_{S1} q_{S1} - T_{SP}2\rho_2 c_{S2}^2\xi w_{P2} - T_{SS}\rho_2 c_{S2} q_{S2} = -\rho_1 c_{S1} q_{S1} \tag{93b}$$

$$R_{SP}c_{P1}\xi + R_{SS}w_{S1} - T_{SP}c_{P2}\xi - T_{SS}w_{S2} = -w_{S1} \tag{93c}$$

$$-R_{SP}w_{P1} + R_{SS}c_{S1}\xi - T_{SP}w_{P2} + T_{SS}c_{S2}\xi = c_{S1}\xi \tag{93d}$$

with $q_{ab} = (1-2c_{ab}^2\xi^2)$ and $w_{ab} = \sqrt{1-c_{ab}^2\xi^2}$, where the index ab can be P1, P2, S1 or S2.

The equation system (93) can also be written in matrix form:

$$\begin{pmatrix} \rho_1 c_{P1} q_{S1} & -2\rho_1 c_{S1}^2\xi w_{S1} & -\rho_2 c_{P2} q_{S2} & 2\rho_2 c_{S2}^2\xi w_{S2} \\ -2\rho_1 c_{S1}^2\xi w_{P1} & -\rho_1 c_{S1} q_{S1} & -2\rho_2 c_{S2}^2\xi w_{P2} & -\rho_2 c_{S2} q_{S2} \\ c_{P1}\xi & w_{S1} & -c_{P2}\xi & -w_{S2} \\ -w_{P1} & c_{S1}\xi & -w_{P2} & c_{S2}\xi \end{pmatrix} \begin{pmatrix} R_{SP} \\ R_{SS} \\ T_{SP} \\ T_{SS} \end{pmatrix} = \begin{pmatrix} 2\rho_1 c_{S1}^2\xi w_{S1} \\ -\rho_1 c_{S1} q_{S1} \\ -w_{S1} \\ c_{S1}\xi \end{pmatrix} \tag{93}$$

The solutions (6) for the reflection and transmission coefficients from the equations systems (88) and (93) are in chapter 2.2.1.

Appendix 2: Layer-wise reconstruction algorithm

One of the major challenges of the material characterization is the determination of the reflection parameters. Especially in a multi-layer structure, the multiple reflections are superimposing the primary echoes, and the reflection parameters can no longer be determined. A possible approach that can still deliver all material and reflections parameters is the layer-wise reconstruction algorithm.

The basic idea is the layer by layer material characterization and the subtraction of the partial reconstructed simulation from the original reflection signal. Theoretically, this removes all multiple reflections of the already determined layers from the reflection signal, and the reflection parameters of the next layer can be determined. The selection of the primary echoes can be done automatically by taking the first echo with a certain amplitude. This approach provides a fully automatic reconstruction and requires a very realistic simulation that contains all sound propagation effects as the measured signal. A more secure approach is user guided echo selection, where it is also possible to ignore artefactual echoes from an inaccurate simulation.

The reconstruction process should be started with a reference measurement to determine the initial amplitude. In addition, the thickness of all layers must be known or determined in advance. The reconstruction process starts with the determination of the surface echo amplitude A_0 and time t_0. By comparison to the surface amplitude of a reference specimen, the longitudinal acoustic impedance $Z_0 = \rho_0 c_{P0}$ of the first layer is determined with equation (55). A simulation is set up with the water delay line and a half infinite material block with the acoustic impedance Z_0. The simulation amplitude is normalized to the measurement and is subtracted from the reflection signal, which now contains everything but the surface echo. Now for each layer n, beginning from the surface, the materials are characterized from a measured reflection signal $s_r(t)$ via the following steps:

1. Deconvolution of $s_r(t)$ to determine the reflection parameters A_{PPn}, t_{PPn}, t_{PSn} of the current layer
2. Determination of the sound velocities with equation (46) and (47) from t_n, t_{PPn}, t_{PSn}
3. Determination of the density from the acoustic impedance and the sound velocity: $\rho_n = Z_n / c_{Pn}$
4. Determination of the acoustic impedance of the next layer from the transmitted amplitude A_T and the backwall echo amplitude: $Z_{n+1} = Z_n \, (A_T + A_{PPn}) \, / \, (A_T - A_{PPn})$

5. Iterative single layer reconstruction: more precise determination of the sound velocities and the density of the current layer and the acoustic impedance of the next layer
6. Subtraction of the new simulation result with the current layer and the half infinite material block of the next layer from the reflection signal
7. Continue with the next layer at 1. or abort

This approach does not provide the ability to determine the ultrasonic attenuation, because the attenuation of the current layer in equation (68) and (69) is not linearly independent from the acoustic impedance of the next layer. If one layer is known, then the attenuation can be determined the same way as for the reconstructions of multiple layers by combining both approaches. The reflection parameters are determined with the layer-wise reconstruction, and the material parameters are determined with the reconstruction of multiple layers at once. Another advantage of this combination is the much better precision. If only the layer-wise algorithm is used, the precision depends very much on the simulation quality and the number of layers. A small error at the first layer produces a very large one at a deeper layer. This effect is avoided by combining both approaches.

This algorithm has not been tested jet.

Appendix 3: Reference specimen 3

The reference specimen 3 (“Ref3”, figure 56) is an eight silicon die stack. It is made for thin and multiple layer investigations with the ultrasonic microscope. Multi-die packages sometimes have delaminations between the single layers. If those can be seen with the ultrasonic microscope, it is often not possible to determine the location properly. For such investigations, artificial air inclusions are brought into the silicon as known delaminations,

All eight 75 µm silicon dies have a 40 µm DAF glue tape laminated on their back side. Before assembly, each die is engraved with a laser at a different position from the active side of the die (figure 57). An overview of all positions and the shape of the air inclusions is shown in figure 58.

The dies with a size of 18.5 mm x 9.5 mm are stacked on a glass substrate manually, without molding and glue curing. This can cause unwanted delaminations. For protection, a 100 µm plastic foil with a 10 µm glue film is manually laminated on the top of the stack. At the position of the air inclusion of the first die it has a small surface dent.

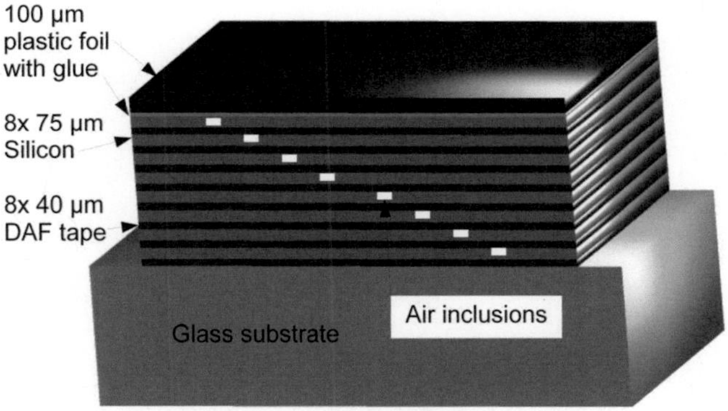

Figure 56: Construction scheme of reference specimen 3 “ref3”

In figure 59 a B-Scan of reference specimen 3 along the B-Scan line (figure 58) is shown. It contains the signals from all eight delaminations of the different layers. An explanation of the single echoes can be found in figure 32. The differences in the reflection signal from a delamination are occurring usually later than the actual defect position is (figure 1). A proper phase inversion can only be detected at the first silicon interface.

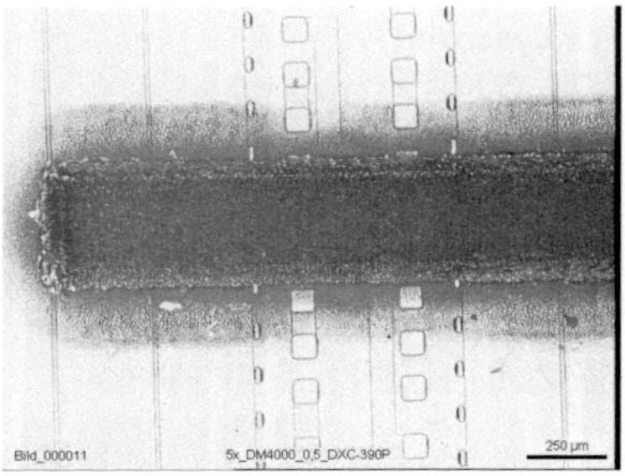

Figure 57: Silicon active side LASER air inclusion before cleaning

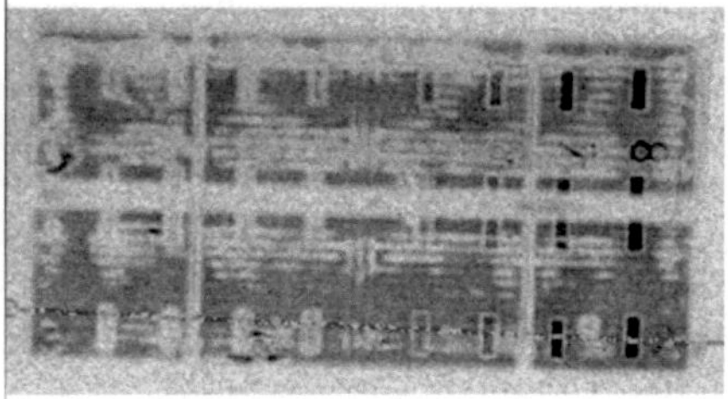

Figure 58: Overview C-Scan of Ref3 with B-Scan line (B-Scan figure 59) through the flaws; scan area: 20.0 mm x 10.5 mm

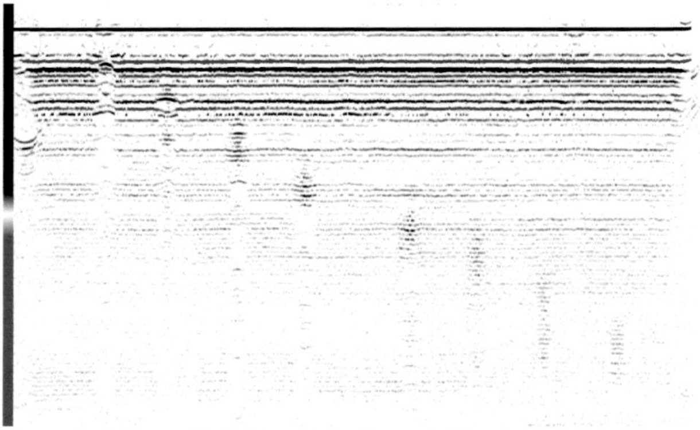

Figure 59: B-Scan of Ref3 through the flaws of all layers deconvoluted data; 75 MHz Transducer; scan area: 18.5 mm x 1.1 µs

The choice of the focus position for such a specimen is very important. For the first layers, focusing on the corresponding layer delivers the best lateral resolution and enough signal amplitude. For deeper layers, the focus position has to be behind the actual layer, to get sufficient signal amplitude from the multiple reflections. The primary echoes are too weak. If the focus position is behind all layers of interest (inside the glass substrate), the information of all layers can be obtained at once – but with reduced lateral resolution for the first layers.

Such a scan, with one deep focus position, has been done to obtain the reflection signals at all positions of the specimen and with enough signal content from all layers. The lateral resolution is not so much of interest here. From this data all C-Scan in figure 60 are generated. They show the artificial air inclusions for each silicon layer and some additional unwanted (semi-)delaminations. The active macroscopic structure of the dies is also visible.

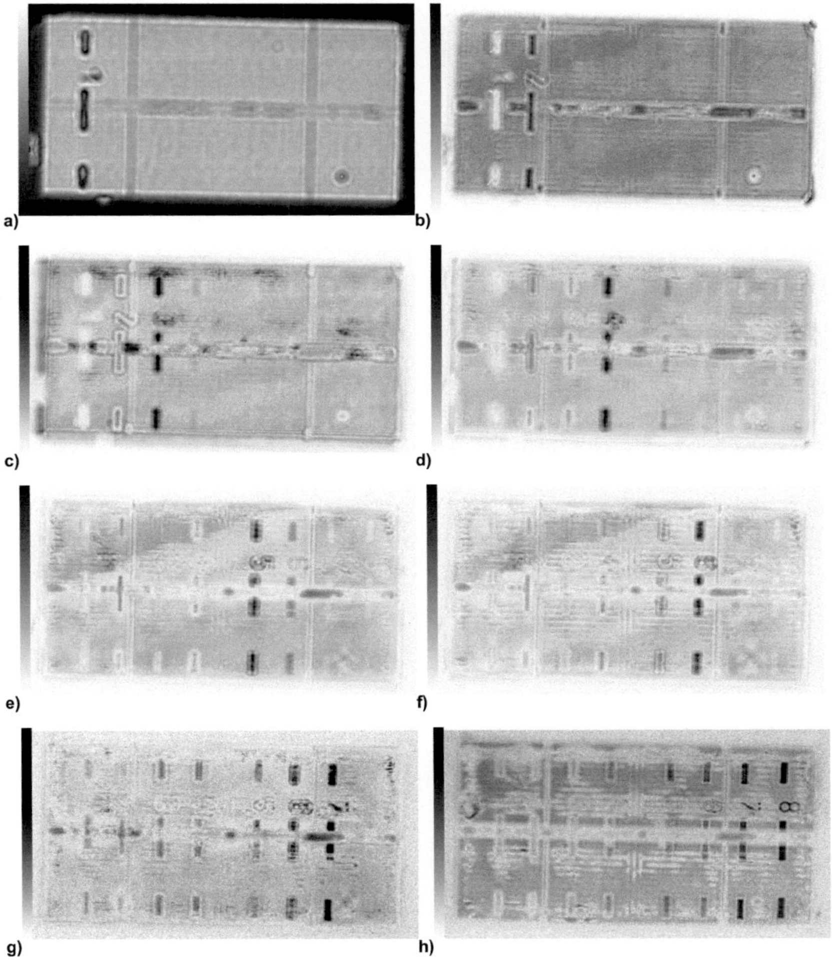

Figure 60: C-Scans of reference specimen 3 from deconvoluted data

75 MHz transducer; focus position behind the 8th silicon layer; all images are from the same data set; scan area: 20.0 mm x 10.5 mm; gray-scale coded integrated absolute amplitudes; a) 1st Si layer, b) 2nd Si layer, c) 3rd Si layer, d) 4th Si layer, e) 5th Si layer, f) 6th Si layer, g) 7th Si layer, h) 8th Si layer

Acknowledgments

I would like to thank all the people, who made this dissertation possible.

Special thanks go to my advisor Professor Elfgard Kühnicke for numerous discussions and constructive criticism. Her experience and depth of knowledge in the field gave me many new impulses. Above all, I would like to thank Professor Klaus J. Wolter for overcoming many bureaucratic hurdles for two collaborations.

This work was also technically and financially supported by our partners Sonoscan and Qimonda.

I thank our partner Sonoscan for making all necessary measurement tasks possible. Working on-site has not only been professionally enriching, it also gave me insight into an internationally leading company for ultrasonic microscopes. I would like to thank Larry Kessler, the chief of Sonoscan. I have greatly enjoyed the uncomplicated collaboration.

The experiments conducted for this work would not have been possible in this quality without the support of Qimonda. The equipment and assistance provided in preparation of the samples were crucial to the success of the dissertation. Especially I would like to mention my supervisor at Qimonda, Dirk Breuer. I thank the staff of the PFA department at Qimonda for their assistance in sample production and preparation.

Finally, I would like to thank my girlfriend and my family for the patience and the sacrifices, which have eased the writing of this dissertation.

Danksagung

Ich möchte all jenen danken, die mir diese Dissertation ermöglicht haben.

Insbesondere möchte ich meiner Betreuerin Prof. Elfgard Kühnicke für viele Diskussionen und konstruktive Kritik danken. Ihre Erfahrung und tiefe Kenntnis ihres Fachgebietes gaben mir viele neue Impulse. Meinem Chef Professor Klaus J. Wolter danke ich vor allem für das Überwinden vieler bürokratischer Hürden um 2 Kooperationen einzugehen.

Diese Arbeit wurde auch durch unsere Kooperationspartner Sonoscan und Qimonda fachlich und finanziell unterstützt.

Unserem Kooperationspartner Sonoscan danke ich für die Ermöglichung der notwendigen Messaufgaben. Die Aufenthalte vor Ort haben mich nicht nur fachlich bereichert, sondern mich auch Einblicke in ein international führendes Unternehmen für Ultraschallmikroskope gegeben. Dem Chef von Sonoscan, Larry Kessler, möchte ich danken. Die unkomplizierte Zusammenarbeit hat mir stets Freude gemacht.

Die für diese Arbeit durchgeführten Experimente wären in dieser Qualität nicht ohne die Unterstützung von Qimonda möglich gewesen. Die zur Verfügung gestellten Geräte und die Unterstützung bei der Probenherstellung waren für das Gelingen der Dissertation entscheidend. Besonders erwähnen möchte ich hier meinen Betreuer bei Qimonda, Dirk Breuer. Den Mitarbeitern der PFA Abteilung bei Qimonda danke ich für ihre Unterstützung bei der Probenherstellung und Präparation.

Abschließend möchte ich noch meiner Freundin und meiner Familie für ihre Geduld und die erbrachten Opfer danken, die mir das Schreiben dieser Dissertation erleichtert haben.

Curriculum vitae

Personal data

	Norbert Gust
Date of birth	6th June 1980
Residence	Dresden, Germany
E-Mail	Norbert.Gust@gmx.de
Languages	German (first), English, French (basics)

School

09/1987 – 07/1993	59. POS Dresden
09/1993 – 07/1999	Gymnasium with Abitur, Dresden Blasewitz (MANOS)

Work experience before Graduation

07/1999 – 09/1999	Fraunhofer IWS Dresden

University

10/1999 – 11/2005	Electrical engineering, Technical University of Dresden emphasis on information technology and acoustics
10/2003 – 01/2004	Applied Physics Laboratory, University of Washington, USA

Work experience

12/2005	Sonoscan Inc., Chicago, USA
01/2006 – 03/2011	Research assistant, TU Dresden, Packaging laboratory
06/2006 – 04/2008	D. candidate at Qimonda Dresden
since 10/2009	Research assistant, TU Dresden, Solid state electronics laboratory